AF517887

Encyclopaedic Dictionary of Telecommunication

Vol. 1

Rajiv Shukla

RAJAT PUBLICATIONS

NEW DELHI - 110 002 (INDIA)

RAJAT PUBLICATIONS
4740/23, Ansari Road, Daryaganj
New Delhi- 110 002 (India)
Phones : +91-11-23267924, +91-9811484973
E-mail: rajat_publication@yahoo.com

Encyclopaedic Dictionary of Telecommunication

Edition 2025

ISBN 978-81-7880-307-4 (Set)

PRINTED IN INDIA

Published by Mrs. Seema Wasan for Rajat Publications, New Delhi and Printed at H.S. Offset Press, Delhi.

Preface

Telecommunications is the transmission of data and information between computers using a communications link such as a standard telephone line. Typically, a basic telecommunications system would consist of a computer or terminal on each end, communication equipment for sending and receiving data, and a communication channel connecting the two users. Appropriate communications software is also necessary to manage the transmission of data between computers. Some applications that rely on this communications technology include the following:

1. Electronic mail (e-mail) is a message transmitted from one person to another through computerized channels. Both the sender and receiver must have access to on-line services if they are not connected to the same network. E-mail is now one of the most frequently used types of telecommunication.

2. Facsimile (fax) equipment transmits a digitized exact image of a document over telephone lines. At the receiving end, the fax machine converts the digitized data back into its original form.

3. Voice mail is similar to an answering machine in that it permits a caller to leave a voice message in a voice mailbox. Messages are digitized so the caller's message can be stored on a disk.

4. Videoconferencing involves the use of computers, television cameras, and communications software and equipment. This equipment makes it possible to conduct electronic meetings while the participants are at different locations.

5. The Internet is a continuously evolving global network of computer networks that facilitates access to information on thousands of topics. The Internet is utilized by millions of people daily.

Actually, telecommunications is not a new concept. It began in the mid-1800s with the telegraph, whereby sounds were translated manually into words; then the telephone, developed in 1876, transmitted voices; and then the teletypewriter, developed in the early 1900s, was able to transmit the written word.

Since the 1960s, telecommunications development has been rapid and wide reaching. The development of dial modem technology accelerated the rate during the 1980s. Facsimile transmission also enjoyed rapid growth during this time. The 1990s have seen the greatest advancement in telecommunications. It is predicted that computing performance will double every eighteen months. In addition, it has been estimated that the power of the computer has doubled thirty-two times since World War II (With row, 1997). The rate of advancement in computer technology shows no signs of slowing.

Deregulation and new technology have created increased competition and widened the range of network services available throughout the world. This increase in telecommunication capabilities allows businesses to benefit from the information revolution in numerous ways, such as streamlining their inventories, increasing productivity, and identifying new markets.

True source for the latest terms and concepts used today in the field of telecommunications. The objective of this book is to provide a self-contained quick-reference to telecommunications jargon and facts in a clear concise manner.

Written and reviewed by industry expert, this dictionary cover topics spanning the telecommunications and information technology industry, including distribution design, insatllation, outside plant, and LAN and internetworking design. The dictionary is based on recognized codes, standards, and regulations, and include extensive glossaries.

—Editor

a. *Abbreviation for* atto (10^{-18}). *See* metric system.

abandoned call. A call in which the call originator disconnects or cancels the call after a connection has been made, but before the call is established.

abbreviated dialing. A telephone service feature that (a) permits the user to dial fewer digits to access a network than are required under the nominal numbering plan, and (b) is limited to a subscriber-selected set of frequently dialed numbers. *Synonym* speed dialing.

abort. 1. In a computer or data transmission system, to terminate, usually in a controlled manner, a processing activity because it is impossible or undesirable for the activity to proceed. 2. In data transmission, a function invoked by a sending station to cause the recipient to discard or ignore all bit sequences transmitted by the sender since the preceding flag sequence.

absolute delay. The time interval or phase difference between transmission and reception of a signal.

absolute temperature. *See* thermodynamic temperature.

absorption band. A spectral region in which the absorption coefficient reaches a relative maximum, by virtue of the physical properties of the matter in which the absorption process takes place.

absorption index. 1. A measure of the attenuation caused by absorption of energy per unit of distance that occurs in an electromagnetic wave of given wavelength propagating in a material medium of given refractive index. *Note:* The value of the absorption index K' is given by the relation

$$K' = \frac{K\lambda}{4\pi n},$$

where K is the absorption coefficient, λ is the wavelength in vacuum, and n is the refractive index of the absorptive material medium. 2. The functional relationship between the Sun angle—at any latitude and local time—and the ionospheric absorption.

absorption loss. That part of the transmission loss caused by the dissipation or conversion of electrical, electromagnetic, or acoustic energy into other forms of energy as a result of its interaction with a material medium.

abstract syntax. In open systems architecture, the specification of application-layer data or application-protocol control information by using notation rules that are independent of the encoding technique used to represent the information.

abstract syntax notation one (ASN.1). A standard, flexible method that (a) describes data structures for representing, encoding, transmitting, and decoding data, (b) provides a set of formal rules for describing the structure of objects independent of machine-specific encoding techniques, (c) is a formal network-management Transmission Control Protocol/Internet Protocol (TCP/IP) language that uses human-readable notation and a compact, encoded representation of the same information used in communications protocols, and (d) is a precise, formal notation that removes ambiguities.

ac. *Abbreviation for* alternating current.

accept. In data transmission, the condition assumed by a primary or secondary station upon correct receipt of a frame for processing.

acceptance. The condition that exists when a system or functional unit meets the specified performance and security requirements.

acceptance pattern. 1. Of an antenna, for a given plane, a distribution plot of the off-axis power relative to the on-axis power as a function of angle or position. *Note:* The acceptance pattern is the equivalent of a horizontal or vertical antenna pattern. 2. Of an optical fiber or fiber bundle, a curve of total transmitted power plotted against the launch angle.

acceptance test. A test of a system or functional unit to ensure that contractual requirements are met. *Note:* An acceptance test may be performed at the factory or user premises by the user, vendor, or a third party.

acceptance testing. Operating and testing of a communication system, subsystem, or component, to ensure that the specified performance characteristics have been met.

acceptance trial. A trial carried out by nominated representatives of the eventual military users of the weapon or equipment to determine if the specified performance and characteristics have been met.

accepted interference. Interference at a higher level than that defined as permissible interference and which has been agreed upon between two or more administrations without prejudice to other administrations.

access. 1. The ability and means necessary to store data in, to retrieve data from, to communicate with, or to make use of any resource of a system. 2. To obtain the use of a resource. 3. (COMSEC) [The] capability and opportunity to gain detailed knowledge of or to alter information or material. 4. (AIS) [The] ability and means to communicate with (*i.e.*, input to or receive output from), or otherwise make use of any information, resource, or component in an AIS. *Note [for 3 and 4]:* An individual does not have "access" if the proper authority or a physical, technical, or procedural measure prevents him/her from obtaining knowledge or having an opportunity to alter information, material, resources, or components. 5. An assigned portion of system resources for one data stream of user communications or signaling.

access attempt. The process by which one or more users interact with a telecommunications system to enable initiation of user information transfer. *Note:* An access attempt begins with an issuance of an access request by an access originator. An access attempt ends either in successful access or in access failure.

access category. A class to which a user, such as a person, program, process, or equipment, of a system may be assigned, based on the resources each user is authorized to use.

access charge. A charge made by a local exchange carrier for use of its local exchange facilities for a purpose such as the origination or termination of traffic that is carried to or from a distant exchange by an interexchange carrier. *Note:* Although some access charges are billed directly to local end users, a very large part of all access charges is paid by interexchange carriers.

access code. The preliminary digits that a user must dial to be connected to a particular outgoing trunk group or line.

access contention. In ISDN applications, *synonymous with "contention" See* contention (def. #2).

access control. 1. A service feature or technique used to permit or deny use of the components of a communication system. 2. A technique used to define or restrict the rights of individuals or application programs to obtain data from, or place data onto, a storage device. 3. The definition or restriction of the rights of individuals or application programs to obtain data from, or place data into, a storage device. 4. [The] process of limiting access to the resources of an AIS to authorized users, programs, processes, or other systems. 5. That function performed by the resource controller that allocates system resources to satisfy user requests.

access control message. A message that is a user request, a resource controller response, or a request/response between resource controllers.

access coupler. *Deprecated term. See* directional coupler.

access-denial time. The time between the start of an access attempt and access failure caused by access denial, *i.e.,* system blocking. *Note:* Access denial times are measured only on access attempts that result in access denial.

access group. A group of one or more stations having identical rights to use the available resources on a PBX, network or host computer.

access line. A transmission path between user terminal equipment and a switching center.

access node. In switching systems, the point where user traffic enters and exits a communications network. *Note:* Access node operations may include various operations, such as protocol conversion and code conversion.

access originator. The functional entity responsible for initiating a particular access attempt. *Note:* An access attempt can be initiated by a source user, a destination user, or the telecommunications system.

access phase. In an information-transfer transaction, the phase during which an access attempt is made. *Note:* The access phase is the first phase of an information-transfer transaction.

access request. A control message issued by an access originator for the purpose of initiating an access attempt.

accounting management. In network management, a set of functions that (a) enables network service use to be measured and the costs for such use to be determined and (b) includes all the resources consumed, the facilities used to collect accounting data, the facilities used to set billing parameters for the services used by customers, maintenance of the data bases used for billing purposes, and the preparation of resource usage and billing reports.

accumulator. 1. A register in which one operand can be stored and subsequently replaced by the result of an arithmetic or logic operation. 2. A storage register. 3. A storage battery.

accuracy. The degree of conformity of a measured or calculated value to its actual or specified value. *Contrast with* precision.

ACD. *Abbreviation for* automatic call distributor.

ACK. *Abbreviation for* acknowledge character.

acknowledge character (ACK). A transmission control character transmitted by the receiving station as an affirmative response to the sending station. *Note:* An acknowledge character may also be used as an accuracy control character.

acknowledgement. 1. A response sent by a receiver to indicate successful receipt of a transmission. *Note:* An example of an acknowledgement is a protocol data unit, or element thereof, between peer entities, to indicate the status of data units that have been successfully received. 2. A message from the addressee informing the originator that the originator's communication has been received and understood.

acknowledgement delay period. *Synonym (loosely)* sliding window.

A-condition. In a start-stop teletypewriter system, the significant condition of the signal element that immediately precedes a character signal or block signal and prepares the receiving equipment for the reception of the code elements. *Contrast with* start signal.

acoustic noise. An undesired audible disturbance in the audio frequency range.

acquisition and tracking orderwire. *See* ATOW.

acquisition time. 1. In a communications system, the time interval required to attain synchronism. 2. In satellite control communications, the time interval required for locking tracking equipment on a signal from a communications satellite.

active satellite. A satellite carrying a station intended to transmit or retransmit radio communication signals. *Note:* An active satellite may perform signal processing functions such as amplification, regeneration, frequency translation, and link switching, to make the signals suitable for retransmission.

active star. *See* star coupler, multiport repeater.

activity factor. For a communications channel during a specified time interval, such as the busy hour, the percentage of time that a signal is present in the channel in either direction.

ACU. *Abbreviation for* automatic calling unit.

A-D. *Abbreviation for* analog-to-digital. *See* analog transmission.

Ada®. The official, high-level computer language of DOD for embedded-computer, real-time applications as defined in MIL-STD-1815. *Note:* Ada® is a registered trademark of the U.S. Government (Ada Joint Program Office).

adaptive channel allocation. In communications system traffic flow control, channel allocation in which information-handling capacities of channels are not predetermined but are assigned on demand. *Note:* Adaptive channel allocation is usually accomplished by means of a multiplexing scheme.

adaptive differential pulse code-modulation (ADPCM). Differential pulse-code modulation in which the prediction algorithm is adjusted in accordance with specific characteristics of the input signal.

adaptive equalization. Equalization (a) that is automatically accomplished while traffic is being transmitted and (b) in which signal characteristics are dynamically adjusted to compensate for changing transmission path characteristics.

adaptive routing. Routing that is automatically adjusted to compensate for network changes such as traffic patterns, channel availability, or equipment failures. *Note:* The experience used for adaptation comes from the traffic being carried.

adaptive system. A system that has a means of monitoring its own performance, a means of varying its own parameters, and uses closed-loop action to improve its performance or to optimize traffic.

ADC. *Abbreviation for* analog-to-digital converter, analog-to-digital conversion.

ADCCP. *Abbreviation for* Advanced Data Communication Control Procedures.

added bit. A bit delivered to the intended destination user in addition to intended user information bits and delivered overhead bits. *Synonym* extra bit.

added block. Any block, or other delimited bit group, delivered to the intended destination user in addition to intended user information bits and delivered overhead bits. *Synonym* extra block.

adder-subtracter. A device that acts as an adder or subtracter depending upon the control signal received; the adder-subtracter may be constructed so as to yield a sum and a difference at the same time. *Note:* An arithmetic adder-subtracter yields arithmetic sums and differences, whereas a logical adder-subtracter yields logical sums and differences.

additive white gaussian noise (AWGN). *Synonym* white noise.

add mode. In addition and subtraction operations, a mode in which the decimal marker is placed at a predetermined location with respect to the last digit entered.

add-on conference. A service feature that allows an additional party to be added to an established call without attendant assistance. *Note:* A common implementation provides a progressive method that allows a call originator or a call receiver to add at least one additional party.

address. 1. In communications, the coded representation of the source or destination of a message. 2. In data processing, a character or group of characters that identifies a register, a particular part of storage, or some other data source or destination. 3. To assign to a device or item of data a label to identify its location. 4. The part of a selection signal that indicates the destination of a call. 5. To refer to a device or data item by its address.

addressability. 1. In computer graphics, the capability of a display surface or storage device to accommodate a specified number of uniquely identifiable points. 2. In micrographics, the capability of a specified field frame to contain a specific number of uniquely identifiable points. *Note:* The addressability is usually specified as the number of identifiable horizontal points by the number of identifiable vertical points, such as 3000 by 4000.

addressable point. In computer graphics, any point of a device that can be addressed.

address field. The portion of a message that contains the source-user address and the destination-user addresses. *Note:* In a communications network, the address field is usually contained within the message header portion of the message. A message usually consists of the message header, the user data, and a trailer.

address-indicating group (AIG). A station or address designator, used to represent a set of specific and frequently recurring combinations of action or information addresses. *Note:* The identity of the message originator may also be included in the AIG. An address group is assigned to each AIG for use as an address designator.

address message. A message sent in the forward direction that contains (a) address information, (b) the signaling information required to route and connect a call to the called line, (c) service-class information, (d) information relating to user and network facilities, and (e) call-originator identity or call-receiver identity.

address message sequencing. In common-channel signaling, a procedure for ensuring that address messages are processed in the correct order when the order in which they are received is incorrect.

address part. A part of an instruction that usually contains only an address or part of an address.

address pattern. A prescribed structure of data used to represent the destination(s) of a block, message, packet, or other formalized data structure.

address resolution protocol (ARP). A Transmission Control Protocol/Internet Protocol (TCP/IP) protocol that dynamically binds a Network-Layer IP address to a Data-Link-Layer physical hardware address, *e.g.,* Ethernet address.

address separator. A character that separates the different addresses in a selection signal.

ADH. *Abbreviation for* automatic data handling.

adjunct service point (ASP). An intelligent-network feature that resides at the intelligent peripheral equipment and responds to service logic interpreter requests for service processing.

administration. 1. Any governmental department or service responsible for discharging the obligations undertaken in the convention of the International Telecommunication Union and the *Regulations.* 2. Internal management of units. 3. The management and execution of all military matters not included in strategy and tactics. 4. In international telecommunications for a given country, the government agency assigned responsibility for the implementation of telecommunications standards, regulations, recommendations, practices, and procedures. 5. In network management, network support functions that ensure that (a) services are performed, (b) the network is used efficiently, and (c) prescribed service-quality objectives are met.

administrative management complex (AMC). In network management, a complex that is controlled by a network provider, and is responsible for and performs network management functions such as network maintenance.

ADP. *Abbreviation for* automatic data processing.

ADPCM. *Abbreviation for* adaptive differential pulse-code modulation.

ADPE. *Abbreviation for* automatic data processing equipment.

ADP system. *Synonym* computer system.

Advanced Data Communication Control Procedures (ADCCP). A bit-oriented Data-Link-Layer protocol used to provide point-to-point and point-to-multipoint transmission of data frames that contain error-control information. *Note:* ADCCP closely resembles high-level data link control (HDLC) and synchronous data link control (SDLC).

advanced intelligent network (AIN). A proposed intelligent-network (IN) architecture that includes both IN/1+ and IN/2 concepts.

advanced television (ATV). A family of television systems that is intended to be improvements over current commercial-quality television. *Note:* The ATV family includes improved-definition television (IDTV), extended-definition television (EDTV), and high-definition television (HDTV).

AECS. *Abbreviation for* Aeronautical Emergency Communications System. *See* Aeronautical Emergency Communications System Plan.

aerial cable. A communications cable designed for installation on, or suspension from, a pole or other overhead structure.

aeronautical advisory station. An aeronautical station used for advisory and civil defense communications primarily with private aircraft stations. *Synonym* UNICOM station.

aeronautical broadcast station. An aeronautical station which makes scheduled broadcasts of meteorological information and notices to airmen. (In certain instances, an aeronautical broadcast station may be placed on board a ship.)

aeronautical Earth station. An Earth station in the fixed-satellite service, or, in some cases, in the aeronautical mobile-satellite service, located at a specified fixed point on land to provide a feeder link for the aeronautical mobile-satellite service.

Aeronautical Emergency Communications System (AECS) Plan. The AECS Plan provides for the operation of aeronautical communications stations, on a voluntary, organized basis, to provide the President and the Federal Government, as well as heads of state and local governments, or their designated representatives, and the aeronautical industry with an expeditious means of communications during an emergency situation.

aeronautical fixed station. A station in the aeronautical fixed service.

aeronautical mobile (OR) [off-route] service. An aeronautical mobile service intended for communications, including those relating to flight coordination, primarily outside national or international civil air routes.

aeronautical mobile (R) [route] service. An aeronautical mobile service reserved for communications relating to safety and regularity of flight, primarily along national or international civil air routes.

aeronautical mobile-satellite service. A mobile-satellite service in which mobile Earth stations are located on board aircraft; survival craft stations and emergency position-indicating radiobeacon stations may also participate in this service.

aeronautical mobile-satellite (OR) [off-route] service. An aeronautical mobile-satellite service intended for communications, including those relating to flight coordination, primarily outside national and international civil air routes.

aeronautical mobile-satellite (R) [route] service. An aeronautical mobile-satellite service reserved for communications relating to safety and regularity of flight, primarily along national or international civil air routes.

aeronautical mobile service. A mobile service between aeronautical stations and aircraft stations, or between aircraft stations, in which survival craft stations may participate; emergency position-indicating radiobeacon stations may also participate in this service on designated distress and emergency frequencies.

aeronautical multicom service. A mobile service not open to public correspondence, used to provide communications essential to conduct activities being performed by or directed from private aircraft.

aeronautical radionavigation-satellite service. A radionavigation-satellite service in which Earth stations are located on board aircraft.

aeronautical radionavigation service. A radionavigation service intended for the benefit and for the safe operation of aircraft.

aeronautical station. A land station in the aeronautical mobile service. In certain instances, an aeronautical station may be located, for example, on board ship or on a platform at sea.

AF. *Abbreviation for* **audio frequency.**

AFNOR. *Acronym for* Association Français Normal. France's national standards-setting organization. *Note:* AFNOR provides the Secretariat for ISO TC97/SC1, Information Technology Vocabulary, which includes computers, communications information processing, and office machines.

AGC. *Abbreviation for* automatic gain control.

AI. *Abbreviation for* artificial intelligence.

AIM. *Abbreviation for* amplitude intensity modulation. *See* intensity modulation.

AIN. *Abbreviation for* advanced intelligent network.

AIOD. *Abbreviation for* automatic identified outward dialing.

AIOD leads. Terminal equipment leads used solely to transmit automatic identified outward dialing (AIOD) data from a PBX to the public switched telephone network or to switched service networks (*e.g.,* EPSCS), so that a vendor can provide a detailed monthly bill identifying long-distance usage by individual PBX stations, tie trunks, or the attendant. . . . [from 47CFR]

air-conditioning. The simultaneous controlling of the characteristics of air, such as temperature, humidity, cleanliness, motion, and pollutant concentration, in a space to meet the requirements of the occupants, a process, or equipment. *Synonym* environmental control.

aircraft Earth station. A mobile Earth station in the aeronautical mobile-satellite service located on board an aircraft.

aircraft station. A mobile station in the aeronautical mobile service, other than a survival craft station, located on board an aircraft.

airdrome control station. An aeronautical station providing communication between an airdrome control tower and aircraft. *Synonym* airport control station.

air-ground radiotelephone service. A public radio service between a base station and airborne mobile stations.

air-ground worldwide communications system. A worldwide military network of ground stations that (a) provides two-way communications links between aircraft and ground stations for navigation and control, including air route traffic control and (b) may also provide support for special functions, such as for civil aircraft providing assistance to military missions and for meeting communications requirements for aircraft flying distinguished visitors.

air portable. Denotes materiel that is suitable for transport by an aircraft loaded internally or externally, with no more than minor dismantling and reassembling within the capabilities of user units. This term must be qualified to show the extent of air portability.

airport control station. *Synonym* airdrome control station.

air sounding. Measuring atmospheric phenomena or determining atmospheric conditions, especially by means of apparatus carried by balloons, rockets, or satellites.

air terminal. In grounding systems, the lightning rod or conductor placed on or above a building, structure, or external conductors for the purpose of intercepting lightning.

AIS. *Abbreviation for* automated information system.

alarm center. A location that receives local and remote alarms. *Note:* An alarm center is usually in a technical control facility.

alarm indicator. A device that responds to a signal from an alarm sensor. *Note:* Examples of alarm indicators include bells, lamps, horns, gongs, and buzzers.

alarm sensor. 1. In communications systems, any device that (a) can sense an abnormal condition within the system and provide a signal indicating the presence or nature of the abnormality to either a local or remote alarm indicator, and (b) may detect events ranging from a simple contact opening or closure to a time-phased automatic shutdown and restart cycle. 2. In a physical security system, an approved device used to indicate a change in the physical environment of a facility or a part thereof. *Note:* Alarm sensors may also be redundant or chained, such as when one alarm sensor is used to protect the housing, cabling, or power protected by another alarm sensor.

a-law. *See* a-law algorithm.

ALE. *Abbreviation for* automatic link establishment.

algorithmic language. An artificial language established for expressing a given class of algorithms.

aligned bundle. A bundle of optical fibers in which the relative spatial coordinates of each fiber are the same at the two ends of the bundle. *Note:* Such a bundle may be used for the transmission of images. *Synonym* coherent bundle.

Allan variance. One half of the time average over the sum of the squares of the differences between successive readings of the frequency deviation sampled over the sampling period. *Note:* The Allan variance is conventionally expressed by $\sigma^2(\tau)$. The samples are taken with no dead-time between them. *Synonym* two-sample variance.

allcall. In adaptive high-frequency (HF) radio automatic link establishment (ALE), a general broadcast that does not request responses and does not designate any specific addresses. *Note:* This essential function is required for emergencies ("HELP"), sounding-type data exchanges, and propagation and connectivity tracking.

all-glass fiber. *Synonym* all-silica fiber.

allocation (of a frequency band). 1. Entry in the Table of Frequency Allocations of a given frequency band for the purpose of its use by one or more (terrestrial or space) radiocommunication services or the radio astronomy service under specified conditions. This term shall also be applied to the frequency band concerned. 2. The process of designating radio-frequency bands for use by specific radio services.

allotment (of a radio frequency or radio frequency channel). Entry of a designated frequency channel in an agreed plan, adopted by a component Conference, for use by one or more administrations for a (terrestrial or space) radiocommunication service in one or more identified countries or geographical areas and under specified conditions.

all trunks busy (ATB). An equipment condition in which all trunks (paths) in a given trunk group are busy. *Note:* All-trunks-busy registers do not indicate subsequent attempts to reach trunk groups.

alphabet. 1. An ordered set of all the letters used in a language, including letters with diacritical signs where appropriate, but not including punctuation marks. 2. An ordered set of all the symbols used in a language, including punctuation marks, numeric digits, nonprinting control characters, and other symbols. *Note:* Examples of alphabets include the Roman alphabet, the Greek alphabet, the Morse Code, and the 128 characters of the American Standard Code for Information Interchange (ASCII).

alphabetic character set. A character set that contains letters and may contain control characters, special characters, and the space character, but not digits.

alphabetic code. A code according to which data are represented through the use of an alphabetic character set.

alphabetic string. 1. A string consisting solely of letters from the same alphabet. 2. A character string consisting solely of letters and associated special characters from the same alphabet.

alphabetic word. 1. A word consisting solely of letters from the same alphabet. 2. A word that consists of letters and associated special characters, but not digits.

alphabet translation. *Deprecated synonym for* alphabet transliteration. *See* alphabet transliteration.

alphabet transliteration. The substitution of the characters of one alphabet for the corresponding characters of a different alphabet, usually accomplished on a character-by-character basis. *Note 1:* An example of alphabet transliteration is the substitution of the Roman letters a, b, and p for the Greek letters α, β, and π, respectively. *Note 2:* Alphabet transliteration is reversible. *Note 3:* Alphabet transliteration often becomes necessary in telecommunications systems because of the different alphabets and codes used worldwide. *Note 4:* In alphabet transliteration, no consideration is given to the meaning of the characters or their combinations.

alphanumeric. 1. Pertaining to a character set that contains letters, digits, and sometimes other characters, such as punctuation marks. 2. Pertaining to a set of unique bit patterns that are used to represent letters of an alphabet, decimal digits, punctuation marks, and other special signs and symbols used in grammar, business, and science, such as those displayed on conventional typewriter keyboards.

alphanumeric character set. A character set that contains both letters and digits, special characters, and the space character.

alphanumeric code. 1. A code derived from an alphanumeric character set. 2. A code that, when used, results in a code set that consists of alphanumeric characters.

alphanumeric data. Data represented by letters, digits, and sometimes by special characters and the space character.

alpha profile. *See* power-law index profile.

altazimuth mount. A mounting, *e.g.*, for a directional antenna, in which slewing takes place in (a) the plane tangent to the surface of the Earth or other frame of reference and (b) elevation about, *i.e.,* above or below, that plane. *Synonym* x-y mount.

alternate mark inversion (AMI) signal. A pseudoternary signal, representing binary digits, in which (a) successive *"marks"* are of alternately positive and negative polarity and the absolute values of their amplitudes are normally equal and (b) *"spaces"* are of zero amplitude.

alternate party. In multilevel precedence and preemption, the call receiver, *i.e.*, the destination user, to which a precedence call will be diverted. *Note 1:* Diversion will occur when the response timer expires, when the call receiver is busy on a call of equal or higher precedence, or when the call receiver is busy with access resources that are non-preemptable. *Note 2:* Alternate party diversion is an optional terminating feature that is subscribed to by the call receiver. Thus, the alternate party is specified by the call receiver at the time of subscription.

alternate routing: The routing of a call or message over a substitute route when a primary route is unavailable for immediate use.

altitude of the apogee or of the perigee. The altitude of the apogee or perigee above a specified reference surface serving to represent the surface of the Earth. *Note:* In technical usage, the definite article is not used with the term *apogee* or *perigee* alone. A body orbiting the Earth is said simply to be *"at apogee"* or *"at perigee."* It may, however, properly be said to be *"at the point of apogee"* or *"at the point of perigee."*

ALU. *Abbreviation for* arithmetic and logic unit.

AM. *Abbreviation for* amplitude modulation.

AMA. *Abbreviation for* automatic message accounting.

amateur-satellite service. A radiocommunication service using space stations on Earth satellites for the same purposes as those of the amateur service.

amateur service. A radiocommunication service for the purpose of self-training, intercommunication and technical investigation carried out by amateurs, that is, by duly authorized persons interested in radio technique solely with a personal aim and without pecuniary interest.

amateur station. A station in the amateur service.

ambient temperature. The temperature of air or other media in a designated area, particularly the area surrounding equipment.

AME. *Abbreviation for* amplitude modulation equivalent, automatic message exchange. *See* compatible sideband transmission.

American National Standards Institute (ANSI). The U.S. standards organization that establishes procedures for the development and coordination of voluntary American National Standards.

American Standard Code for Information Interchange (ASCII). *See* ASCII.

AMI. *Abbreviation for* alternate mark inversion. *See* alternate mark inversion signal.

AMI violation. A *"mark"* that has the same polarity as the previous *"mark"* in the transmission of alternate mark inversion (AMI) signals. *Note:* In some transmission protocols, AMI violations are deliberately introduced to facilitate synchronization or to signal a special event.

amplifier. *See* fiber amplifier, optical repeater.

amplitude equalizer. A corrective network that is designed to modify the amplitude characteristics of a circuit or system over a desired frequency range. *Note:* Such devices may be fixed, manually adjustable, or automatic.

amplitude hit. *See* hit (def. #1).

amplitude intensity modulation (AIM). *Deprecated term. See* intensity modulation.

amplitude keying. Keying in which the amplitude of a signal is varied among the members of a set of discrete values.

amplitude modulation equivalent (AME). *Synonym* compatible sideband transmission.

amplitude-vs.-frequency distortion. Distortion in a transmission system caused by nonuniform attenuation, or gain, in the system with respect to frequency under specified operating conditions. *Synonym* frequency distortion.

AMPS. *Abbreviation for* automatic message processing system.

AMTS. *Abbreviation for* automated maritime telecommunications system.

analog computer. A device that performs operations on data that are represented within the device by continuous variables having a physical resemblance to the quantities being represented. *Note:* The earliest analog computers were constructed with purely mechanical components, such as levers, cogs, cams, discs, and gears. These components represented the quantities being manipulated or the operator-inserted values. Modern analog computers usually employ electrical parameters, such as voltages, resistances, or currents to represent the quantities being manipulated.

analog data. Data represented by a physical quantity that is considered to be continuously variable and has a magnitude directly proportional to the data or to a suitable function of the data.

analog facsimile equipment. Facsimile equipment in which (a) analog techniques are used to encode the image detected by the scanner and (b) the output is an analog signal. *Note:* Examples of analog facsimile equipment are CCITT Group 1 and CCITT Group 2 facsimile equipment.

analog switch. Switching equipment designed, designated, or used to connect circuits between users for real-time transmission of analog signals.

analog-to-digital (A-D) coder. *Synonym* analog-to-digital converter (ADC).

analog-to-digital converter (ADC). A device that converts an analog signal to a digital signal that represents equivalent information. *Synonyms* analog-to-digital (A-D) coder, analog-to-digital (A-D) encoder.

analog-to-digital (A-D) encoder. *Synonym* analog-to-digital converter (ADC).

analog transmission. Transmission of a continuously varying signal as opposed to transmission of a discretely varying signal.

angle of deviation. In optics, the net angular deflection experienced by a light ray after one or more refractions or reflections.

angle of incidence. The angle between an incident ray and the normal to a reflecting or refracting surface.

angstrom (Å). A unit of length equal to 10^{-10} m. *Note 1:* The angstrom is not an SI (International System) unit, and it is not accepted for government use (Fed. Std. 376B). *Note 2:* The angstrom is, and historically has been, used in the fields of optics, spectroscopy, and microscopy.

ANI. *Abbreviation for* automatic number identification.

anisochronous. Pertaining to transmission in which the time interval separating any two significant instants in sequential signals is not necessarily related to the time interval separating any other two significant instants. *Note:* Isochronous and anisochronous are characteristics, while synchronous and asynchronous are relationships.

anisochronous transmission. *See* asynchronous transmission.

anomalous propagation (AP). Abnormal propagation caused by fluctuations in the properties (such as density and refractive index) of the propagation medium. *Note:* AP may result in the reception of signals well beyond the distances usually expected.

ANS. Abbreviation for American National Standard.

ANSI. *Abbreviation for* American National Standards Institute.

ANSI/EIA/TIA-568. A U.S. industry standard that specifies a generic telecommunications cabling system, which will support a multiproduct, multivendor environment, for commercial buildings. *Note 1:* The standard specifies performance characteristics for unshielded twisted pair telecommunications cabling, including categories allowing data communications up to 100 Mb/s. These categories are designated 3, 4, and 5. Categories 1 and 2 have not been defined. *Note 2:* The standard has been adopted as FIPS PUB 174.

answer back. A signal sent by receiving equipment to the sending station to indicate that the receiver is ready to accept transmission.

answer signal. A supervisory signal returned from the called telephone to the originating switch when the call receiver answers. *Note 1:* The answer signal stops the ringback signal from being returned to the caller. *Note 2:* The answer signal is returned by means of a closed loop.

antenna. Any structure or device used to collect or radiate electromagnetic waves.

antenna aperture. *See* aperture (def #1).

antenna array. An assembly of antenna elements with dimensions, spacing, and illumination sequence such that the fields for the individual elements combine to produce a maximum intensity in a particular direction and minimum field intensities in other directions.

antenna blind cone. The volume of space, usually approximately conical with its vertex at the antenna, that cannot be scanned by an antenna because of limitations of the antenna radiation pattern and mount. *Note:* An example of an antenna blind cone is that of an air route surveillance radar (ARSR). The horizontal radiation pattern of an ARSR antenna is very narrow. The vertical radiation pattern is fan-shaped, reaching approximately 70° of elevation above the horizontal plane. As the antenna is rotated about a vertical axis, it can illuminate targets only if they are 70° or less from the horizontal plane. Above that elevation, they are in the antenna blind cone. *Synonym* cone of silence.

antenna coupler. A device used to match the impedance of a transmitter and/or receiver to an antenna to provide maximum power transfer.

antenna dissipative loss. A power loss resulting from changes in the measurable impedance of a practical antenna from a value theoretically calculated for a perfect antenna.

antenna gain-to-noise-temperature (G/T). In the characterization of antenna performance, a figure of merit, where G is the antenna gain in decibels at the receive frequency, and T is the equivalent noise temperature of the receiving system in kelvins.

antenna height above average terrain. The antenna height above the average terrain elevations from 3.2 to 16 kilometers (2 to 10 miles) from the antenna for the eight directions spaced evenly for each 45° of azimuth starting with true north. *Note:* In general, a different antenna height above average terrain will be determined in each direction from the antenna. The average of these eight heights is the antenna height above average terrain. In some cases, such as seashore, fewer than eight directions may be used.

antenna lobe. A three-dimensional section of the radiation pattern of a directional antenna, bounded by one or more cones of nulls or by regions of diminished irradiance.

antenna matching. The process of adjusting impedance so that the input impedance of an antenna equals or approximates the characteristic impedance of its transmission line over a specified range of frequencies. *Note:* The impedance of either the transmission line, or the antenna, or both, may be adjusted to effect the match.

anti-clockwise polarized wave. *Synonym* left-hand polarized wave.

anti-interference. Pertaining to equipment, processes, or techniques used to reduce the effect of natural and man-made noise on radio communications.

anti-jam. Measures to ensure that intended transmitted information can be received despite deliberate jamming attempts.

anti-node. A point in a standing wave at which the amplitude is a maximum.

anti-spoof. Measures to prevent an opponent's participation in a telecommunications network, or operation/control of a cryptographic or COMSEC system.

anycall. In adaptive high-frequency (HF) radio automatic link establishment, a broadcast in which (a) the called stations are unspecified, (b) stations receiving the call stop scanning, and (c) each station automatically responds in pseudorandom time slots.

AP. *Abbreviation for* anomalous propagation.

APC. *Abbreviation for* adaptive predictive coding.

APD. *Abbreviation for* avalanche photodiode. *Note:* apd and a.p.d. are also used.

aperiodic antenna. An antenna designed to have an approximately constant input impedance over a wide range of frequencies. *Note:* Examples of aperiodic antennas include terminated rhombic antennas and wave antennas. *Synonym* nonresonant antenna.

aperture. 1. In a directional antenna, the portion of a plane surface very near the antenna normal to the direction of maximum radiant intensity, through which the major part of the radiation passes. 2. In an acoustic device that launches a sound wave, the passageway, determined by the size of a hole in the inelastic material and the wavelength.

aperture distortion. In facsimile, the distortion of the recorded image caused by the shape and finite size of the scanning and recording apertures. *Note:* The distortion may occur in one or more attributes of the recorded image, such as in resolution, density, or shape.

aperture illumination. 1. The field distribution, in amplitude and phase, over the antenna physical aperture. 2. The phase and amplitude of the element feed voltages or the distribution of the currents in an array of elements.

aperture-to-medium coupling loss. The difference between the theoretical gain of a very large antenna, such as the antennas in beyond-the-horizon microwave links, and the gain that can be realized in operation. *Note 1:* Aperture-to-medium coupling loss is related to the ratio of the scatter angle to the antenna beamwidth. *Note 2:* The "very large antennas" are referred to in wavelengths; thus, this loss can apply to line-of-sight systems also.

API. *Abbreviation for* application program interface.

apogee. In an orbit of a satellite orbiting the Earth, the point that is farthest from the gravitational center of the Earth.

apogee altitude. *See* altitude of the apogee or of the perigee.

Application Layer. *See* Open Systems Interconnection–Reference Model.

application program interface (API). A formalized set of software calls and routines that can be referenced by an application program in order to access supporting network services.

applique. Circuit components added to an existing system to provide additional or alternate functions. *Note:* Applique may be used to modify carrier telephone equipment designed for ringdown manual operation to allow for use between points having dial equipment.

approved circuit. *Deprecated synonym for* protected distribution system.

aramid yarn. Generic name for a tough synthetic yarn that is often used in optical cable construction for the strength member, protective braid, and/or rip cord for jacket removal.

Archie. Distributed-system-based software that searches indices of files available on public servers on the Internet. *Note 1: Archie* servers may provide access via telnet, E-mail, or a special Archie client. *Note 2:* Using Archie requires a user to be familiar with where the indices are located, *i.e.*, the user must provide an Archie server domain name or IP address.

architecture. *See* computer architecture, network architecture.

archiving. The storing of files, records, and other data for reference and alternative backup.

area broadcast shift. The changing from listening to transmissions intended for one broadcast area to listening to transmissions intended for another broadcast area. *Note 1:* An area broadcast shift may occur when a ship or aircraft crosses the boundary between listening areas. *Note 2:* Shift times, on the date a ship or aircraft is expected to pass into another area, must be strictly observed or the ship or aircraft will miss messages intended for it. *Synonym* radio watch shift.

area code. *See* access code, code, country code, NXX code.

*****argument.** 1. An independent variable. 2. Any value of an independent variable. *Note:* Examples of arguments include search keys, numbers that identify the location of a data item in a table, and the è in sin è.

arithmetic and logic unit (ALU). A part of a computer that performs arithmetic, logic, and related operations.

arithmetic overflow. 1. In a digital computer, the condition that occurs when a calculation produces a result that is greater than a given register or storage location can store or represent. 2. In a digital computer, the amount that a calculated value is greater than a given register or storage location can store or represent. *Note:* The overflow may be placed at another location. *Synonym* overflow.

arithmetic register. A register that holds the operands or the results of operations such as arithmetic operations, logic operations, and shifts.

arithmetic shift. A shift, applied to the representation of a number in a fixed radix numeration system and in a fixed-point representation system, and in which only the characters representing the fixed-point part of the number are moved. An arithmetic shift is usually equivalent to multiplying the number by a positive or a negative integral power of the radix, except for the effect of any rounding; compare the logical shift with the arithmetic shift, especially in the case of floating-point representation.

arithmetic underflow. In a digital computer, the condition that occurs when a calculation produces a non-zero result that is less than the smallest non-zero quantity that a given register or storage location can store or represent.

arithmetic unit. In a processor, the part that performs arithmetic operations; sometimes the unit performs both arithmetic and logic operations.

Armed Forces Radio Service (AFRS). A radio broadcasting service that is operated by and for the personnel of the armed services in the area covered by the broadcast. *Note:* An example of an AFRS is the radio service operated by the U.S. Army for U.S. and allied military personnel on duty in overseas areas.

ARP. *Abbreviation for* address resolution protocol.

ARPANET. *Abbreviation for* Advanced Research Projects Agency Network. A packet-switching network used by the Department of Defense, later evolved into the Internet.

ARQ. *Abbreviation for* automatic repeat-request. Error control for data transmission in which the receiver detects transmission errors in a message and automatically requests a retransmission from the transmitter. *Note:* Usually, when the transmitter receives the ARQ, the transmitter retransmits the message until it is either correctly received or the error persists beyond a predetermined number of retransmissions. *Synonyms* error-detecting-and-feedback system, repeat-request system.

array. 1. An arrangement of elements in one or more dimensions. 2. In a programming language, an aggregate that consists of data objects with identical attributes, each of which may be uniquely referenced by subscription.

array processor. A processor capable of executing instructions in which the operands may be arrays rather than data elements. *Synonym* vector processor.

arrester. A device that protects hardware, such as systems, subsystems, circuits, and equipment, from voltage or current surges produced by lightning or electromagnetic pulses. *Note:* If the hardware is adequately protected, associated software may also be adequately protected. *Synonym* surge suppressor.

ARS. *Abbreviation for* automatic route selection.

articulation index. A measure of the intelligibility of voice signals, expressed as a percentage of speech units that are understood by the listener when heard out of context. *Note:* The articulation index is affected by noise, interference, and distortion.

articulation score (AS). A subjective measure of the intelligibility of a voice system in terms of the percentage of words correctly understood over a channel perturbed by interference. *Note:* Articulation scores have been experimentally obtained as functions of varying word content, bandwidth, audio signal-to-noise ratio and the experience of the talkers and listeners involved.

artificial intelligence (AI). The capability of a device to perform functions that are normally associated with human intelligence, such as reasoning and optimization through experience. *Note:* AI is the branch of computer science that attempts to approximate the results of human reasoning by organizing and manipulating factual and heuristic knowledge. Areas of AI activity include expert systems, natural language understanding, speech recognition, vision, and robotics.

artificial transmission line. A four-terminal electrical network, *i.e.* an electrical circuit, that has the characteristic impedance, transmission time delay, phase shift, and/or other parameter(s) of a real transmission line and therefore can be used to simulate a real transmission line in one or more of these respects. *Colloquial synonym* art line.

art line. *Colloquial synonym for* artificial transmission line.

ARU. *Abbreviation for* audio response unit.

ASCII. *Acronym for* American Standard Code for Information Interchange. The standard code used for information interchange among data processing systems, data communications systems, and associated equipment in the United States. *Note 1:* The ASCII character set contains 128 coded characters. *Note 2:* Each ASCII character is a 7-bit coded unique character; 8 bits when a parity check bit is included. *Note 3:* The ASCII character set consists of control characters and graphic characters. *Note 4:* When considered simply as a set of 128 unique bit patterns, or 256 with a parity bit, disassociated from the character equivalences in national implementations, the ASCII may be considered as an alphabet used in machine languages. *Note 5:* The ASCII is the U.S. implementation of International Alphabet No. 5 (IA No. 5) as specified in CCITT Recommendation V.3.

ASP. *Abbreviation for* adjunct service point.

aspect ratio. In facsimile or television, the ratio of the width to the height of a picture, document, or scanning field.

assemble. To translate a computer program expressed in an assembly language into a machine language.

assembler. A computer program that is used to assemble. *Synonym* assembly program.

assembly. In logistics, an item forming a portion of an equipment that can be provisioned and replaced as an entity and which normally incorporates replaceable parts or groups of parts.

assembly language. A computer-oriented language (a) in which instructions are symbolic and usually in one-to-one correspondence with sets of machine language instructions and (b) that may provide other facilities, such as the use of macro instructions. *Synonym* computer-dependent language.

assembly program. *Synonym* assembler.

assembly time. The elapsed time taken for the execution of an assembler.

assignment. For NS/EP, the designation of priority level(s).

assignment (of a radio frequency or radio frequency channel). Authorization given by an administration for a radio station to use a radio frequency or radio frequency channel under specified conditions.

associative storage. 1. A storage device whose storage locations are identified by their contents, or by a part of their contents, rather than by their names or positions. *Note:* Associative storage can also refer to this process as well as to the device. *Synonym* content-addressable storage. 2. Storage that supplements another storage.

asymmetrical modulator. *Synonym* unbalanced modulator.

asynchronous communications system. A data communications system that uses asynchronous operation. *Note 1:* In an asynchronous communications system, extra signal elements are usually appended to the data for the purpose of synchronizing individual data characters or blocks. *Note 2:* The time spacing between successive data characters or blocks may be of arbitrary duration. *Synonym* start-stop system.

asynchronous network. A network in which the clocks do not need to be synchronous or mesochronous. *Synonym* nonsynchronous network.

asynchronous operation. 1. A sequence of operations in which operations are executed out of time coincidence with any event. 2. An operation that occurs without a regular or predictable time relationship to a specified event; *e.g.,* the calling of an error diagnostic routine that may receive control at any time during the execution of a computer program. *Synonym* asynchronous working.

asynchronous time-division multiplexing (ATDM). Time-division multiplexing in which asynchronous transmission is used.

asynchronous transmission. Data transmission in which the instant that each character, or block of characters, starts is arbitrary; once started, the time of occurrence of each signal representing a bit within the character, or block, has the same relationship to significant instants of a fixed time frame.

asynchronous working. *Synonym* asynchronous operation.

ATB. *Abbreviation for* all trunks busy.

AT Commands. A *de facto* standard for modem commands from an attached CPU, used in most 1,200 and 2,400 b/s modems.

ATDM. *Abbreviation for* asynchronous time-division multiplexing.

ATM. *Abbreviation for* asynchronous transfer mode.

atmospheric noise. Radio noise caused by natural atmospheric processes, primarily lightning discharges in thunderstorms.

atomic time. *See* International Atomic Time.

ATOW. *Acronym for* acquisition and tracking orderwire. A downlink circuit that provides a terminal with information regarding uplink acquisition and synchronization status.

attachment unit interface (AUI). In a local area network, the interface between the medium access unit (MAU) and the data terminal equipment within a data station.

attack time. The time between (a) the instant that a signal at the input of a device or circuit exceeds the activation threshold of the device or circuit and (b) the instant that the device or circuit reacts in a specified manner, or to a specified degree, to the input. *Note:* Attack time occurs in devices such as clippers, peak limiters, compressors, and voxes.

attendant access loop. A switched circuit that provides an attendant with a manual means for call completion and control. *Note:* An attendant access loop might be given a specific telephone number.

attendant conference. A network-provided service feature that allows an attendant to establish a conference connection of three or more users.

attendant position. The part of a switching system used by an attendant, *i.e.,* an operator, to assist users in call completion and use of special services.

attenuation coefficient. The rate of diminution of average power with respect to distance along a transmission path. *Synonym* attenuation rate.

attenuation constant. 1. The real part of the propagation constant in any electromagnetic propagation medium. *Note 1:* The attenuation constant is usually expressed as a numerical value per unit length. *Note 2:* The attenuation constant may be calculated or experimentally determined for each medium. 2. For a particular propagation mode in an optical fiber, the real part of the axial propagation constant.

attenuation-limited operation. The condition that prevails when attenuation, rather than bandwidth, limits the performance of a communications link.

attenuation rate. *Synonym* attenuation coefficient.

attribute. 1. In database management, a property inherent in an entity or associated with that entity for database purposes. 2. In network management, a property of a managed object that has a value. *Note 1:* Mandatory initial values for attributes can be specified as part of the managed object class definition. *Note 2:* Attributes may be either mandatory or conditional.

ATV. *Abbreviation for* advanced television.

audible. *Synonym for* audible ringing tone.

audible ringing tone. In telephony, a signal, usually consisting of an audio tone interrupted at a slow repetition rate, provided to a caller to indicate that the called-party instrument is being sent a ringing signal. *Note:* The audible ringing tone may be generated by the called-party servicing switch or by the calling-party servicing switch, but it is not generated by the called telephone instrument. *Synonyms* audible, ringback tone.

audio frequency (AF). The band of frequencies (approximately 20 Hz to 20 kHz) that, when transmitted as acoustic waves, can be heard by the normal human ear.

audit. 1. To conduct an independent review and examination of system records and activities in order to test the adequacy and effectiveness of data security and data integrity procedures, to ensure compliance with established policy and operational procedures, and to recommend any necessary changes. 2. Independent review and examination of records and activities to assess the adequacy of system controls, to ensure compliance with established policies and operational procedures, and to recommend necessary changes in controls, policies, or procedures.

audit review file. A file created by executing statements included in a computer program for the explicit purpose of providing data for auditing.

audit trail. 1. A record of both completed and attempted accesses and service. 2. Data in the form of a logical path linking a sequence of events, used to trace the transactions that have affected the contents of a record. 3. Chronological record of system activities to enable the reconstruction and examination of the sequence of events and/or changes in an event. *Note:* Audit trail may apply to information in an AIS, or to the transfer of COMSEC material.

AUI. *Abbreviation for* attachment unit interface.

aurora. Sporadic radiant emission from the upper atmosphere that usually occurs about the North and South magnetic poles of the Earth. *Note 1:* Auroras are most intense at times of intense magnetic storms caused by sunspot activity. The distribution of auroral intensity with altitude shows a pronounced maximum near 100 km above the Earth. Auroras may occasionally be observed within 40° or less of the equator. *Note 2:* Auroras interfere with radio communications. *Note 3:* In the Northern hemisphere, the aurora is called the Aurora Borealis (Northern Lights). In the Southern hemisphere, the aurora is called the Aurora Australis (Southern Lights).

authenticate. 1. To establish, usually by challenge and response, that a transmission attempt is authorized and valid. 2. Verify the identity of a user, user device, or other entity, or the integrity of data stored, transmitted, or otherwise exposed to unauthorized modification in an automated information system, or establish the validity of a transmitted message. 3. A challenge given by voice or electrical means to attest to the authenticity of a message or transmission.

authentication. [Any] Security measure designed to establish the validity of a transmission, message, or originator, or a means of verifying an individual's eligibility to receive specific categories of information.

authenticator. 1. A symbol or group of symbols, or a series of bits, selected or derived in a prearranged manner and usually inserted at a predetermined point within a message or transmission for the purpose of attesting to the validity of the message or transmission. 2. A letter, numeral, group of letters or numerals, or any combinations of these, attesting to the authenticity of a message or transmission. 3. [In INFOSEC,] means used to confirm the identity or eligibility of a station, originator, or individual.

authorization. 1. The rights granted to a user to access, read, modify, insert, or delete certain data, or to execute certain programs. 2. Access rights granted to a user, program, or process.

authorized bandwidth. Authorized bandwidth is, for purposes of this Manual, the necessary bandwidth (bandwidth required for transmission and reception of intelligence) and does not include allowance for transmitter drift or Doppler shift.

authorized frequency. A frequency that is allocated and assigned by a competent authority to a specific user for a specific purpose. *See* assigned frequency.

AUTODIN. *Acronym for* automatic digital network. *See* Defense Data Network.

automated data medium. *Synonym* machine-readable medium.

automated information system (AIS). 1. An assembly of computer hardware, software, firmware, or any combination of these, configured to accomplish specific information-handling operations, such as communication, computation, dissemination, processing, and storage of information. 2. [In INFOSEC,] any equipment or interconnected system or subsystems of equipment that is used in the automatic acquisition, storage, manipulation, management, movement, control, display, switching, interchange, transmission or reception of data and includes computer software, firmware, and hardware. *Note:* Included are computers, word processing systems, networks, or other electronic information handling systems, and associated equipment.

automated information systems security. 1. Measures and controls that ensure confidentiality, integrity, and availability of the information processed and stored by automated information systems. *Note 1:* The unauthorized disclosure, modification, or destruction may be accidental or intentional. *Note 2:* Automated information systems security includes consideration of all hardware and software functions, characteristics and features; operational procedures; accountability procedures; and access controls at the central computer facility, remote computer, and terminal facilities; management constraints; physical structures and devices, such as computers, transmission lines, and power sources; and personnel and communications controls needed to provide an acceptable level of risk for the automated information system and for the data and information contained in the system. Automated information systems security also includes the totality of security safeguards needed to provide an acceptable protection level for an automated information system and for the data handled by an automated information system. 2. In INFOSEC, *synonym* computer security.

automated maritime telecommunications system (AMTS). An automatic, integrated and interconnected maritime communications system serving ship stations on specified inland and coastal waters of the United States.

automated radio. A radio that can be automatically controlled by electronic devices and that requires little or no human intervention.

automated tactical command and control system. A command and control system, or part thereof, that manipulates the movement of information from source to user without human intervention. *Note:* In an automated tactical command and control system, automated execution of a decision without human intervention is not mandatory.

automatic answering. A service feature in which the called terminal automatically responds to the calling signal and the call may be established whether or not the called terminal is attended by an operator.

automatic callback. A service feature that permits a user, when encountering a busy condition, to instruct the system to retain the called and calling numbers and to establish the call when there is an available line. *Note 1:* Automatic callback may be implemented in the terminal, in the switching system, or shared between them. *Note 2:* Automatic callback is not the same as camp-on.

automatic call distributor (ACD). A device that distributes incoming calls to a specific group of terminals. *Note:* If the number of active calls is less than the number of terminals, the next call will be routed to the terminal that has been in the idle state the longest. If all terminals are busy, the incoming calls are held in a first-in-first-out queue until a terminal becomes available.

automatic calling. Calling in which the elements of the selection signal are entered into the data network contiguously at the full data signaling rate. The selection signal is generated by the data terminal equipment. *Note:* A limit may be imposed by the design criteria of the network to prevent more than a permitted number of unsuccessful call attempts to the same address within a specified period.

automatic calling unit (ACU). A device that enables equipment, such as computers and card dialers, to originate calls automatically over a telecommunications network.

automatic data handling (ADH). 1. A generalization of automatic data processing to include the aspect of data transfer. 2. Combining data processing and data transfer.

automatic data processing (ADP). 1. An interacting assembly of procedures, processes, methods, personnel, and equipment to perform automatically a series of data processing operations on data. *Note:* The data processing operations may result in a change in the semantic content of the data. 2. Data processing by means of one or more devices that use common storage for all or part of a computer program, and also for all or part of the data necessary for execution of the program; that execute user-written or user-designated programs; that perform user-designated symbol manipulation, such as arithmetic operations, logic operation, or character-string manipulations; and that can execute programs that modify themselves during their execution. *Note:* Automatic data processing may be performed by a stand-alone unit or by several connected units. 3. Data processing largely performed by automatic means. [JP1] 4. That branch of science and technology concerned with methods and techniques relating to data processing largely performed by automatic means.

automatic data processing equipment (ADPE). Any equipment or interconnected system or subsystems of equipment that is used in the automatic acquisition, storage, manipulation, management, movement, control, display, switching, interchange, transmission, or reception, of data or information (i) by a Federal agency, or (ii) under a contract with a Federal agency which (i) requires the use of such equipment, or (ii) requires the performance of a service or the furnishing of a product which is performed or produced making significant use of such equipment. Such term includes (i) computer, (ii) ancillary equipment, (iii) software, firmware, and similar procedures, (iv) services, including support services, and (v) related resources as defined by regulations issued by the Administrator for General Services. . . . [Public Law 99-500, Title VII, Sec. 822 (a) Section 111(a) of the *Federal Property and Administrative Services Act of 1949 (40 U.S.C. 759(a)) revised.*]

automatic dialing. *See* automatic calling unit.

Automatic Digital Network (AUTODIN). Formerly, a worldwide data communications network of the Defense Communications System, now replaced by the Defense Switched Network (DSN).

automatic error correction. *See* error-correcting code.

automatic exchange. In a telephone system, an exchange in which communications among users are effected by means of switches set in operation by the originating user equipment without human intervention at the central office or branch exchange.

automatic function. A machine function or series of machine functions controlled by a program and carried out without assistance of an operator.

automatic gain control (AGC). A process or means by which gain is automatically adjusted in a specified manner as a function of a specified parameter, such as received signal level.

automatic identified outward dialing (AIOD). A service feature of some switching or terminal devices that provides the user with an itemized statement of usage on directly dialed calls. *Note:* AIOD is facilitated by automatic number identification (ANI) equipment to provide automatic message accounting (AMA).

automatic message accounting (AMA). A service feature that automatically records data regarding user-dialed calls.

automatic message exchange (AME). In an adaptive high-frequency (HF) radio network, an automated process allowing the transfer of a message from message injection to addressee reception, without human intervention. *Note:* Through the use of machine-addressable transport guidance information, *i.e.,* the message header, the message is automatically routed through an on-line direct connection through single or multiple transmission media.

automatic message processing system (AMPS). Any organized assembly of resources and methods used to collect, process, and distribute messages largely by automatic means.

automatic number identification (ANI). A service feature in which the directory number or equipment number of a calling station is automatically obtained. *Note:* ANI is used in message accounting.

automatic operation. The functioning of systems, equipment, or processes in a desired manner at the proper time under control of mechanical or electronic devices that operate without human intervention.

automatic redial: A service feature that allows the user to dial, by depressing a single key or a few keys, the most recent telephone number dialed at that instrument. *Note:* Automatic redial is often associated with the telephone instrument, but may be provided by a PBX, or by the central office. *Synonym* last number redial. *Contrast with* automatic calling unit.

automatic reload. *See* bootstrap (def.#3).

automatic remote rekeying. [In INFOSEC, a] procedure to rekey a distant crypto-equipment electronically without specific actions by the receiving terminal operator. *Note:* Automatic remote rekeying may also apply to non-crypto devices.

automatic remote reprogramming and rekeying. The procedure by which distant equipment is reprogrammed or rekeyed electronically without specific actions by the receiving terminal.

automatic repeat-request (ARQ). *See* ARQ.

automatic ringdown circuit. A circuit providing priority telephone service, typically for key personnel; the circuit is activated when the telephone handset is removed from the cradle causing a ringing signal to be sent to the distant unit(s). *See* verified off-hook.

automatic route selection (ARS). Electronic or mechanical selection and routing of outgoing calls without human intervention.

Automatic Secure Voice Communications Network (AUTOSEVOCOM). A worldwide, switched, secure voice network developed to fulfill DOD long-haul, secure voice requirements.

automatic sequential connection. A service feature in which the terminals at each of a set of specified addresses are automatically connected, in a predetermined sequence, to a single terminal at a specified address.

automatic signaling service. *Synonym* hotline.

automatic sounding. The testing of selected channels or paths by providing a very brief beacon-like identifying broadcast that may be used by other stations to evaluate connectivity, propagation, and availability, and to identify known working channels for possible later use for communications or calling. *Note 1:* Automatic soundings are primarily intended to increase the efficiency of the automatic link establishment (ALE) function, thereby increasing system throughput. *Note 2:* Sounding information is used for identifying the specific channel to be used for a particular ALE connectivity attempt.

automatic switching system. 1. In data communications, a switching system in which all the operations required to execute the three phases of information-transfer transactions are automatically executed in response to signals from a user end-instrument. *Note:* In an automatic switching system, the information-transfer transaction is performed without human intervention, except for initiation of the access phase and the disengagement phase by a user. 2. In telephony, a system in which all the operations required to set up, supervise, and release connections required for calls are automatically performed in response to signals from a calling device.

Automatic Voice Network (AUTOVON). Formerly, the principal long-haul, unsecure voice communications network within the Defense Communications System, now replaced by the Defense Switched Network (DSN).

automation. 1. The implementation of processes by automatic means. 2. The investigation, design, development, and application of methods of rendering processes automatic, self-moving, or self-controlling. 3. The conversion of a procedure, a process, or equipment to automatic operation.

AUTOSEVOCOM. *Acronym for* Automatic Secure Voice Communications Network.

AUTOVON. *Acronym for* Automatic Voice Network. *Superseded by* Defense Switched Network.

auxiliary operation. An offline operation performed by equipment not under control of the processing unit.

auxiliary storage. 1. Storage that is available to a processor only through its input/output channels. 2. In a computer, any storage that is not internal memory, *i.e.,* is not random access memory (RAM). *Note:* Examples of auxiliary storage media are magnetic diskettes, optical disks including CD ROM, and magnetic tape cassettes.

availability. 1. The degree to which a system, subsystem, or equipment is operable and in a committable state at the start of a mission, when the mission is called for at an unknown, *i.e.,* a random, time. *Note 1:* The conditions determining operability and committability must be specified. *Note 2:* Expressed mathematically, availability is 1 minus the unavailability. 2. The ratio of (a) the total time a functional unit is capable of being used during a given interval to (b) the length of the interval. *Note 1:* An example of availability is 100/168 if the unit is capable of being used for 100 hours in a week. *Note 2:* Typical availability objectives are specified in decimal fractions, such as 0.9998.

available line. 1. In voice, video, or data communications, a circuit between two points that is ready for service, but is in the idle state. 2. In facsimile transmission, the portion of the scanning line that can be specifically used for image signals. *Synonym* useful line.

available time. From the point of view of a user, the time during which a functional unit can be used. *Note:* From the point of view of operating and maintenance personnel, the available time is the same as the uptime, *i.e.,* the time during which a functional unit is fully operational.

avalanche photodiode (APD). A photodiode that operates with a reverse-bias voltage that causes the primary photocurrent to undergo amplification by cumulative multiplication of charge carriers. *Note:* As the reverse-bias voltage increases toward the breakdown, hole-electron pairs are created by absorbed photons. An avalanche effect occurs when the hole-electron pairs acquire sufficient energy to create additional pairs when the incident photons collide with the ions, *i.e.,* the holes and electrons. Thus, a signal gain is achieved.

average picture level (APL). In video systems, the average level of the picture signal during active scanning time integrated over a frame period; defined as a percentage of the range between blanking and reference white level.

average rate of transmission. *Synonym* effective transmission rate.

avoidance routing. The assignment of a circuit path to avoid certain critical or trouble-prone circuit nodes.

AWGN. *Abbreviation for* additive white gaussian noise. *See* white noise.

axial propagation constant. In an optical fiber, the propagation constant evaluated along the optical axis of the fiber in the direction of transmission. *Note:* The real part of the axial propagation constant is the attenuation constant. The imaginary part is the phase constant.

axial ratio. Of an electromagnetic wave having elliptical polarization, the ratio of the magnitudes of the major axis and the minor axis of the ellipse described by the electric field vector.

axial ray. A light ray that travels along the optical axis.

b. *Abbreviation for* bit.

B. *Abbreviation for* bel, byte.

babble. In transmission systems, the aggregate of crosstalk induced in a given line by all other lines.

backbone. 1. The high-traffic-density connectivity portion of any communications network. 2. In packet-switched networks, a primary forward-direction path traced sequentially through two or more major relay or switching stations. *Note:* In packet-switched networks, a backbone consists primarily of switches and interswitch trunks.

background noise. The total system noise in the absence of information transmission.

background processing. The execution of lower priority computer programs when higher priority programs are not using the system resources. *Note:* Priorities may be assigned by system software, application software, or the operator.

backup. *See* backup file.

backup file. A copy of a file made for purposes of later reconstruction of the file, if necessary. *Note:* A backup file may be used for preserving the integrity of the original file and may be recorded on any suitable medium. *Synonym* job-recovery control file.

backward recovery. The reconstruction of an earlier version of a file by using a newer version of data recorded in a journal.

backward signal. A signal sent from the called to the calling station, *i.e.,* from the original data sink to the original data source. *Note:* Backward signals are usually sent via a backward channel and may consist of supervisory, acknowledgment, or control signals.

backward supervision. The use of supervisory signal sequences from a secondary to a primary station.

balance. In electrical circuits and networks, to adjust the impedance to achieve specific objectives, such as to reach specified return loss objectives at a hybrid junction of two-wire and four-wire circuits.

balanced. Pertaining to electrical symmetry.

balanced signal pair. *Synonym* balanced line.

balance return loss. 1. A measure of the degree of balance between two impedances connected to two conjugate sides of a hybrid set, coil, network, or junction. 2.. A measure of the effectiveness with which a balancing network simulates the impedance of a two-wire circuit at a hybrid coil.

balun. *Abbreviation for* balanced to unbalanced. In radio frequency usage, a device used to couple a balanced device or line to an unbalanced device or line.

band. 1. In communications, the frequency spectrum between two defined limits. 2. A group of tracks on a magnetic drum or on one side of a magnetic disk. 3. A set of frequencies authorized for use in a geographical area defined for common carriers for purposes of communications system management.

band-elimination filter. *Synonym* band-stop filter.

bandpass limiter. A device that imposes hard limiting on a signal and contains a filter that suppresses the unwanted products (harmonics) of the limiting process.

band-rejection filter. *Synonym* band-stop filter.

band-suppression filter. *Synonym* band-stop filter.

bandwidth balancing mechanism. In a distributed-queue dual-bus network, a procedure in which a node occasionally skips the use of empty queued arbitrated slots, and which procedure effects sharing of the bandwidth mechanisms.

bandwidth compression. 1. The reduction of the bandwidth needed to transmit a given amount of data in a given time. 2. The reduction of the time needed to transmit a given amount of data in a given bandwidth. *Note:* Bandwidth compression implies a reduction in normal bandwidth of an information-carrying signal without reducing the information content of the signal.

bandwidth•distance product. Of an optical fiber, under specified launching and cabling conditions, at a specified wavelength, a figure of merit equal to the product of the fiber's length and the 3-dB bandwidth of the optical signal. *Note 1:* The bandwidth•distance product is usually stated in megahertz•kilometer (MHz•km) or gigahertz•kilometer (GHz•km). *Note 2:* The bandwidth•distance product, which is normalized to 1 km, is a useful figure of merit for predicting the effective fiber bandwidth for other lengths, and for concatenated fibers. *Synonym* bandwidth•length product.

bandwidth•length product. *Synonym for* bandwidth•distance product.

bandwidth-limited operation. The condition prevailing when the system bandwidth limits performance. *Note:* Bandwidth-limited operation occurs when the system distorts the signal waveform beyond specified limits. For linear systems, bandwidth-limited operation is equivalent to distortion-limited operation.

bar code. A code representing characters by sets of parallel bars of varying thickness and separation that are read optically by transverse scanning. *Note:* Bar code uses include identifying merchandise, sorting mail, and inventorying supplies.

barrage jamming. Jamming accomplished by transmitting a band of frequencies that is large with respect to the bandwidth of a single emitter. *Note:* Barrage jamming may be accomplished by presetting multiple jammers on adjacent frequencies or by using a single wideband transmitter. Barrage jamming makes it possible to jam emitters on different frequencies simultaneously and reduces the need for operator assistance or complex control equipment. These advantages are gained at the expense of reduced jamming power at any given frequency.

base. 1. In the numeration system commonly used in scientific notation, the real number that is raised to a power denoted by the exponent and then multiplied by the coefficient to determine the value of the number represented without the use of exponents. *Note:* An example of a base is the number 6.25 in the expression $2.70 \times 6.25^{1.5} \approx 42.19$. The 2.70 is the coefficient and the 1.5 is the exponent. In the decimal numeration system, the base is 10 and in the binary numeration system, the base is 2. The value $e \approx 2.718$ is the natural base. 2. A reference value. 3. A number that is multiplied by itself as many times as indicated by an exponent.

baseband local area network. A local area network in which information is encoded, multiplexed, and transmitted without modulation of carriers.

baseband modulation. Intensity modulation of an optical source, *e.g.*, LED or ILD, directly, without first modulating the signal of interest onto an electrical carrier wave.

baseband signaling. Transmission of a digital or analog signal at its original frequencies; *i.e.*, a signal in its original form, not changed by modulation.

basecom. *Abbreviation for* base communications.

base communications (basecom). Communications services, such as the installation, operation, maintenance, augmentation, modification, and rehabilitation of communications networks, systems, facilities, and equipment, including off-post extensions, provided for the operation of a military post, camp, installation, station, or activity. *Synonym* communications base station.

base Earth station. An Earth station in the fixed-satellite service or, in some cases, in the land mobile-satellite service, located at a specified fixed point or within a specified area on land to provide a feeder link for the land mobile-satellite service.

base station. 1. A land station in the land mobile service. 2. In personal communication service, the common name for all the radio equipment located at one fixed location, and that is used for serving one or several cells.

basic exchange telecommunications radio service (BETRS). A commercial service that can extend telephone service to rural areas by replacing the local loop with radio communications. *Note:* In the BETRS, non-government ultra-high frequency (UHF) and very high frequency (VHF) common carrier and the private radio service frequencies are shared.

basic group. *See* group.

basic mode link control. Control of data links by use of the control characters of the 7-bit character set for information processing interchange as given in ISO Standard 646-1983 and CCITT Recommendation V.3-1972.

basic rate interface (BRI). A CCITT Integrated Services Digital Network (ISDN) multipurpose user interface standard that denotes the capability of simultaneous voice and data services provided over two clear 64-kb/s channels and one clear 16-kb/s channel (2B+D) access arrangement to each user location.

basic service. 1. A pure transmission capability over a communication path that is virtually transparent in terms of its interaction with customer-supplied information. 2. The offering of transmission capacity between two or more points suitable for a user's transmission needs and subject only to the technical parameters of fidelity and distortion criteria, or other conditioning.

basic service element (BSE). 1. An optional unbundled feature, generally associated with the basic serving arrangement (BSA), that an enhanced-service provider (ESP) may require or find useful in configuring an enhanced service. 2. A fundamental (basic) communication network service; an optional network capability associated with a BSA. *Note:* BSEs constitute optional capabilities to which the customer may subscribe or decline to subscribe.

basic serving arrangement (BSA). 1. The fundamental tariffed switching and transmission (and other) services that an operating company must provide to an enhanced service provider (ESP) to connect with its customers through the company network. 2. In an open-network-architecture context, the fundamental underlying connection of an enhanced service provider (ESP) to and through the operating company's network including an ESP access link, the features and functions associated with that access link at the central office serving the ESP and/or other offices, and the transport (dedicated or switched) within the network that completes the connection from the ESP to the central office serving its customers or to capabilities associated with the customer's complementary network services. *Note:* Each component may have a number of categories of network characteristics. Within these categories of network characteristics are alternatives from among which the customer must choose. Examples of BSA components are ESP access link, transport and/or usage.

basic status. In data transmission, the status of the capability of a secondary station to send or receive a frame containing an information field.

batched communications. *Synonym* batched transmission.

batched transmission. The transmission of two or more messages from one station to another without intervening responses from the receiving station. *Synonym* batched communications.

batch processing. 1. The processing of data or the accomplishment of jobs accumulated in advance in such a manner that the user cannot further influence the processing while it is in progress. 2. The processing of data accumulated over a period of time. 3. *Loosely,* the execution of computer programs serially. 4. Pertaining to the technique of executing a set of computer programs such that each is completed before the next program of the set is started. 5. Pertaining to the sequential input of computer programs or data.

Baudot code. A synchronous code in which five equal-length bits represent one character. *Note 1:* The Baudot code, which was developed circa 1880, has been replaced by the start-stop asynchronous International Alphabet No. 2 (IA No. 2). *Note 2:* IA No. 2 is not, and should not be identified as, the Baudot code. *Note 3:* The Baudot code has been widely used in teletypewriter systems.

BCC. *Abbreviation for* block check character.

BCD. *Abbreviation for* binary coded decimal.

B channel. 1. A communications channel used for the transmission of an aggregate signal generated by multichannel transmitting equipment. 2. The CCITT designation for a clear channel, 64-kb/s service capability provided to a subscriber under the Integrated Services Digital Network offering. *Note:* The B channel, also called the bearer channel, is intended for transport of user information, as opposed to signaling information.

BCH code. *Abbreviation for* Bose-Chaudhuri-Hochquenghem code. A multilevel, cyclic, error-correcting, variable-length digital code used to correct errors up to approximately 25% of the total number of digits. *Note:* BCH codes are not limited to binary codes, but may be used with multilevel phase-shift keying whenever the number of levels is a prime number or a power of a prime number, such as 2, 3, 4, 5, 7, 8, 11, and 13. A BCH code in 11 levels has been used to represent the 10 decimal digits plus a sign digit.

BCI. *Abbreviation for* bit-count integrity.

Bd. *Abbreviation for* baud.

beacon. *See* radiobeacon station.

beam. 1. The main lobe of an antenna radiation pattern. 2. A column of light. *Note:* A beam may be parallel, divergent, or convergent.

beam divergence. Of an electromagnetic beam, in any plane that intersects the beam axis, the increase in beam diameter with distance from the aperture from which the beam emerges. *Note 1:* Beam divergence is usually used to characterize electromagnetic beams in the optical regime, *i.e.,* cases in which the aperture from which the beam emerges is very large with respect to the wavelength. *Note 2:* Beam divergence usually refers to a beam of circular cross section, but not necessarily so. A beam may, for example, have an elliptical cross section, in which case the orientation of the beam divergence must be specified, *e.g.,* with respect to the major or minor axis of the elliptical cross section.

beamsplitter. A device for dividing an optical beam into two or more separate beams. *Note:* An example of a beamsplitter is a partially reflecting mirror.

beamwidth. 1. In the radio regime, of an antenna pattern, the angle between the half-power (3-dB) points of the main lobe, when referenced to the peak effective radiated power of the main lobe. *Note:* Beamwidth is usually expressed in degrees. It is usually expressed for the horizontal plane, but may also be expressed for the vertical plane. 2. For the optical regime, *see* beam divergence.

bearer channel. *See* B channel.

bearer service. A telecommunications service that allows transmission of user-information signals between user-network interfaces. *See* B channel, service access.

beating. *See* heterodyne.

beeping. *Synonym* paging, radio paging.

B8ZS. *Abbreviation for* bipolar with eight-zero substitution. A T-carrier line code in which bipolar violations are deliberately inserted if user data contains a string of 8 or more consecutive zeros. *Note 1:* B8ZS is used to ensure a sufficient number of transitions to maintain system synchronization when the user data stream contains an insufficient number of "ones" to do so. *Note 2:* B8ZS is used in the European hierarchy at the T1 rate.

bell (BEL) character. A transmission control character that is used when there is a need to call for user or operator attention in a communications system, and that usually activates an audio or visual alarm or other attention-getting device.

Bell Operating Company (BOC). Any of the 22 operating companies that were divested from AT&T by court order. *Note:* Cincinnati Bell Telephone Co. and Southern New England Bell Telephone Co. are not included.

bend loss. *See* macrobend loss, microbend loss.

BER. *Abbreviation for* bit error ratio.

BERT. *Acronym for* bit error ratio tester.

BETRS. *Abbreviation for* basic exchange telecommunications radio service.

between-the-lines entry. Unauthorized access to a momentarily inactive terminal, of a legitimate user, assigned to a communications channel.

BEX. *Abbreviation for* broadband exchange.

bias distortion. 1. Signal distortion resulting from a shift in the bias. 2. In binary signaling, distortion of the signal in which all the significant intervals have uniformly longer or shorter durations than their theoretical durations. *Note:* Bias distortion is expressed in percent of the system-specified unit interval.

biconical antenna. An antenna consisting of two conical conductors, having a common axis and vertex, and extending in opposite directions. *Note 1:* In a biconical antenna, excitation is applied at the common vertex. *Note 2:* If one of the cones is reduced to a plane, the antenna is called a discone.

bidirectional asymmetry. In data transmission, the condition that exists when information flow characteristics are different in each direction.

bidirectional coupler. *See* directional coupler.

bidirectional symmetry. The condition that exists when information flow characteristics are the same in each direction.

BIH. *French abbreviation for* International Time Bureau. *See* International Atomic Time.

bilateral control. *Synonym* bilateral synchronization.

bilateral synchronization. A synchronization control system between exchanges A and B in which the clock at exchange A controls the data received at exchange B and the clock at exchange B controls the data received at exchange A. *Note:* Bilateral synchronization is usually implemented by deriving the timing from the incoming bit stream. *Synonym* bilateral control.

binary. 1. Pertaining to a selection, choice, or condition that has two possible different values or states. 2. Pertaining to a fixed radix numeration system that has a radix of 2.

binary code. A code, the elements of which can assume either one of two possible states.

binary-coded decimal (BCD). Pertaining to the representation of a decimal digit by a unique arrangement of no fewer than four binary digits.

binary-coded decimal code. *Synonym* binary-coded decimal notation.

binary-coded decimal interchange code. *See* binary-coded decimal notation.

binary-coded decimal (BCD) notation. A binary notation in which each of the decimal digits is represented by a binary numeral. *Synonyms* binary-coded decimal code, binary-coded decimal representation.

binary-coded decimal representation. *Synonym* binary-coded decimal notation.

binary digit (bit). *See* bit.

binary element. A constituent element of data that takes either of two values or states. *Note: "Binary element"* should not be confused with *"binary digit".*

binary exponential backoff. *See* truncated binary exponential backoff.

binary modulation. The process of varying a parameter of a carrier as a function of two finite, discrete states.

binary notation. 1. Any notation that uses two different characters, usually the digits 0 and 1. *Note:* Data encoded in binary notation need not be in the form of a pure binary numeration system; *e.g.,* they may be represented by a Gray code. 2. A scheme for representing numbers, which scheme is characterized by the arrangements of digits in sequence, with the understanding that successive digits are interpreted as coefficients of successive powers of base 2.

binary number. A number that is expressed in binary notation and is usually characterized by the arrangement of bits in sequence, with the understanding that successive bits are interpreted as coefficients of successive powers of the base 2.

binary synchronous (bi-sync) communication. A character-oriented, data-link-layer protocol. *Note:* The bi-sync protocol is being phased out of most computer communication networks in favor of bit-oriented protocols such as SDLC, HDLC, and ADCCP.

binding. In computer, communications, and automatic data processing systems, assigning a value or referent to an identifier. *Note:* Examples of binding include assigning a value to a parameter, assigning an absolute address to a virtual or relative address, and assigning a device identifier to a symbolic address or label.

biphase modulation. *Synonym* phase-shift keying.

bipolar signal. A signal that may assume either of two polarities, neither of which is zero. *Note 1:* A bipolar signal may have a two-state non-return-to-zero (NRZ) or a three-state return-to-zero (RZ) binary coding scheme. *Note 2:* A bipolar signal is usually symmetrical with respect to zero amplitude, *i.e.,* the absolute values of the positive and negative signal states are nominally equal.

birefringence. In a transparent material, anisotropism of the refractive index, which varies as a function of polarization as well as orientation with respect to the incident ray. *Note 1:* The term *"birefringence"* means, literally, *"double refraction."* *Note 2:* All crystals except those of cubic lattice structure exhibit some degree of anisotropy with regard to their physical properties, including refractive index. Other materials, such as glasses or plastics, become birefringent when subjected to mechanical strain. *Note 3:* Birefringent materials, including crystals, have the ability to refract an unpolarized incident ray into two separate, orthogonally polarized rays, which in the general case take different paths, depending on orientation of the material with respect to the incident ray. The refracted rays are referred to as the "ordinary," or "O" ray, which obeys Snell's Law, and the "extraordinary," or "E" ray, which does not. *Synonym* double refraction.

birefringent medium. *See* birefringence.

B-ISDN. *Abbreviation for* broadband ISDN.

bistable. Pertaining to a device capable of assuming either one of two stable states.

bistable multivibrator. *Synonym* flip-flop.

bistable trigger circuit. *Synonym* flip-flop.

bi-sync. *Abbreviation for* binary synchronous (communication).

bit. *Abbreviation for* binary digit. 1. A character used to represent one of the two digits in the numeration system with a base of two, and only two, possible states of a physical entity or system. 2. In binary notation either of the characters 0 or 1. 3. A unit of information equal to one binary decision or the designation of one of two possible and equally likely states of anything used to store or convey information.

bit-by-bit asynchronous operation. In data transmission, an operation in which manual, semiautomatic, or automatic shifts in the data modulation rate are accomplished by gating or slewing the clock modulation rate. *Note:* For example, bit-by-bit asynchronous operation may be at 50 b/s one moment and at 1200 b/s the next moment.

bit configuration. The sequence of bits used to encode a character.

bit-count integrity (BCI). 1. In message communications, the preservation of the exact number of bits that are in the original message. 2. In connection-oriented services, preservation of the number of bits per unit time. *Note:* Bit-count integrity is not the same as bit integrity, which requires that the delivered bits correspond exactly with the original bits.

bit density. The number of bits recorded per unit length, area, or volume. *Note:* Bit density is the reciprocal of bit pitch. *Synonym* recording density.

biternary transmission. Digital transmission in which two binary pulse trains are combined for transmission over a channel in which the available bandwidth is sufficient for transmission of only one of the two pulse trains at a time if they remain in binary form.

bit error rate. *Deprecated term. See* bit error ratio.

bit error ratio (BER). The number of erroneous bits divided by the total number of bits transmitted, received, or processed over some stipulated period. *Note 1:* Examples of bit error ratio are (a) transmission BER, *i.e.,* the number of erroneous bits received divided by the total number of bits transmitted; and (b) information BER, *i.e.,* the number of erroneous decoded (corrected) bits divided by the total number of decoded (corrected) bits. *Note 2:* The BER is usually expressed as a coefficient and a power of 10; for example, 2.5 erroneous bits out of 100,000 bits transmitted would be 2.5 out of 10^5 or 2.5 x 10^{-5}.

bit error ratio tester (BERT). A testing device that compares a received data pattern with a known transmitted pattern to determine the level of transmission quality.

bit interval. *See* bit, character interval, unit interval.

bit inversion. 1. The changing of the state of a bit to the opposite state. 2. The changing of the state that represents a given bit, *i.e.,* a 0 or a 1, to the opposite state. *Note:* For example, if a 1 is represented by a given polarity or phase at one stage in a circuit, the 1 is represented by the opposite polarity or phase at the next stage.

bit pairing. The practice of establishing, within a code set, a number of subsets that have an identical bit representation except for the state of a specified bit. *Note:* An example of bit pairing occurs in the International Alphabet No. 5 and the American Standard Code for Information Interchange (ASCII), where the upper case letters are related to their respective lower case letters by the state of bit six.

bit position. A character position in a word in a binary notation.

bit rate (BR). In a bit stream, the number of bits occurring per unit time, usually expressed in bits per second. *Note:* For n-ary operation, the bit rate is equal to $\log_2 n$ times the rate (in bauds), where *n* is the number of significant conditions in the signal.

bit-rate•distance product. *See* bandwidth•distance product.

bit robbing. In digital carrier systems, the practice or technique of preempting, at regular intervals and for the purpose of transmitting signaling information, one digit time slot that (a) is associated with the given user channel for which signaling is required, and (b) is used primarily for transporting encoded speech via that channel. *Note 1:* Bit robbing is an option in networks compatible with T-carrier, *e.g.,* an ISDN. *Note 2:* In conventional T-carrier systems, bit robbing uses, in every sixth frame, the time slot associated with the least significant bit. *Synonym* speech digit signaling.

bit-sequence independence. A characteristic of some digital data transmission systems that impose no restrictions on, or modification of, the transmitted bit sequence. *Note:* Bit-sequence-independent protocols are in contrast to protocols that reserve certain bit sequences for special meanings, such as the flag sequence, 01111110, for HDLC, SDLC, and ADCCP protocols.

bit slip. In digital transmission, the loss of a bit or bits, caused by variations in the respective clock rates of the transmitting and receiving devices. *Note:* One cause of bit slippage is overflow of a receive buffer that occurs when the transmitter's clock rate exceeds that of the receiver. This causes one or more bits to be dropped for lack of storage capacity.

bits per inch (b/in). A unit used to express the linear bit density of data in storage. *Note:* The abbreviation "*bpi*" is not in accordance with international standards, and is therefore deprecated.

bits per second (b/s). A unit used to express the number of bits passing a designated point per second. *Note 1:* For example, for two-condition serial transmission in a single channel in which each significant condition represents a bit, *i.e.*, a 0 or a 1, the bit rate in bits per second and the baud have the same numerical value only if each bit occurs in a unit interval. In this case, the data signaling rate in bits per second is 1/T, where T is the unit interval. *Note 2:* The abbreviation "*bps*" is not in accordance with international standards, and is therefore deprecated.

bit-stepped. Pertaining to the control of digital equipment in which operations are performed one step at a time at the applicable bit rate.

bit-stream transmission. 1. In bit-oriented systems, the transmission of bit strings. 2. In character-oriented systems, the transmission of bit streams that represent characters. *Note:* In bit-stream transmission, the bits usually occur at fixed time intervals, start and stop signals are not used, and the bit patterns follow each other in sequence without interruption.

bit string. A sequence of bits. *Note:* In a bit stream, individual bit strings may be separated by data delimiters.

bit stuffing. The insertion of noninformation bits into data. *Note 1:* Stuffed bits should not be confused with overhead bits. *Note 2:* In data transmission, bit stuffing is used for various purposes, such as for synchronizing bit streams that do not necessarily have the same or rationally related bit rates, or to fill buffers or frames. The location of the stuffing bits is communicated to the receiving end of the data link, where these extra bits are removed to return the bit streams to their original bit rates or form. Bit stuffing may be used to synchronize several channels before multiplexing or to rate-match two single channels to each other. *Synonym* positive justification.

bit stuffing rate. *See* nominal bit stuffing rate.

bit synchronization. Synchronization in which the decision instant is brought into alignment with the received bit, *i.e.,* the basic signaling element.

bit synchronous operation. Operation in which data circuit terminating equipment (DCE), data terminal equipment (DTE), and transmitting circuits are all operated in bit synchronism with a clock. *Note 1:* In bit synchronous operation, clock timing is usually delivered at twice the modulation rate, and one bit is transmitted or received during each clock cycle. *Note*

2: Bit synchronous operation is sometimes erroneously referred to as digital synchronization.

BIU. *Abbreviation for* bus interface unit. *See* network interface device.

BLACK. [A] designation applied to telecommunications and automated information systems, and to associated areas, circuits, components, and equipment, in which only unclassified signals are processed. *Note:* Encrypted signals are unclassified.

blackbody. A totally absorbing body that does not reflect radiation. *Note:* In thermal equilibrium, a blackbody absorbs and radiates at the same rate; the radiation will just equal absorption when thermal equilibrium is maintained.

black facsimile transmission. 1. In facsimile systems using amplitude modulation, that form of transmission in which the maximum transmitted power corresponds to the maximum density of the subject. 2. In facsimile systems using frequency modulation, that form of transmission in which the lowest transmitted frequency corresponds to the maximum density of the subject.

black noise. Noise that has a frequency spectrum of predominately zero power level over all frequencies except for a few narrow bands or spikes. *Note:* An example of black noise in a facsimile transmission system is the spectrum that might be obtained when scanning a black area in which there are a few random white spots. Thus, in the time domain, a few random pulses occur while scanning.

black recording. 1. In facsimile systems using amplitude modulation, recording in which the maximum received power corresponds to the maximum density of the record medium. 2. In a facsimile system using frequency modulation, recording in which the lowest received frequency corresponds to the maximum density of the record medium.

black signal. In facsimile, the signal resulting from scanning a maximum-density area of the object.

BLACK signal. A signal that represents only unclassified or encrypted information, usually in cryptographic systems.

blanketing. The interference that is caused by the presence of an AM broadcast signal of one volt per meter (V/m) or greater strengths in the area adjacent to the antenna of the transmitting station. The 1 V/m contour is referred to as the blanket contour and the area within this contour is referred to as the "*blanket area.*"

blanketing area. In the vicinity of a transmitting antenna, the area in which the signal from that antenna interferes with the reception of other signals. *Note:* The blanketing area around a given transmitting antenna depends on the selectivity and sensitivity of the receiver, and on the respective levels of the other signals in question.

blanking. In graphic display, the suppression of the display of one or more display elements or display segments.

blind transmission. Transmission without obtaining a receipt, *i.e.,* acknowledgement of reception, from the intended receiving station. *Note:* Blind transmission may occur or be necessary when security constraints, such as radio silence, are imposed, when technical difficulties with a sender's receiver or receiver's transmitter occur, or when lack of time precludes the delay caused by waiting for receipts.

blinking. In graphic display devices, an intentional periodic change in the intensity of one or more display elements or display segments.

block. 1. A group of bits or digits that is transmitted as a unit and that may be encoded for error-control purposes. 2. A string of records, words, or characters, that for technical or logical purposes are treated as a unit. *Note 1:* Blocks (a) are separated by interblock gaps, (b) are delimited by an end-of-block signal, and (c) may contain one or more records. *Note 2:* A block is usually subjected to some type of block processing, such as multidimensional parity checking, associated with it. 3. In programming languages, a subdivision of a program that serves to group related statements, delimit routines, specify storage allocation, delineate the applicability of labels, or segment parts of the program for other purposes.

block character. *See* end-of-transmission-block character.

block check. In the processing or transmission of digital data, an error-control procedure that is used to determine whether a block of data is structured according to given rules.

block check character (BCC). A character added to a transmission block to facilitate error detection. *Note:* In longitudinal redundancy checking and cyclic redundancy checking, block check characters are computed for, and added to, each message block transmitted. This block check character is compared with a second block check character computed by the receiver to determine whether the transmission is error free.

block code. An error detection and/or correction code in which the encoded block consists of N symbols, containing K information symbols ($K<N$) and N-K redundant check symbols, such that most naturally occurring errors can be detected and/or corrected.

block diagram. A diagram of a system, a computer, or a device in which the principal parts are represented by suitably annotated geometrical figures to show both the basic functions of the parts and their functional relationships.

block distortion. In the received image in video systems, distortion characterized by the appearance of an underlying block encoding structure.

block efficiency. In a block, the ratio of the number of user information bits to the total number of bits. *Note:* For a given block scheme, block efficiency represents the maximum possible efficiency for a given block scheme transmitted over a perfect transmission link.

block-error probability. The expected block-error ratio.

block-error ratio. The ratio of the number of incorrectly received blocks to the total number of blocks transferred. *Note:* The block-error ratio is calculated using empirical measurements. Multiple block-error ratios may be used to predict block-error probability.

blocking. 1. The formatting of data into blocks for purposes of transmission, storage, checking, or other functions. 2. Denying access to, or use of, a facility, system, or component. 3. The failure of a telecommunications network to meet a user service demand, because of the lack of an available communications path.

blocking criterion. In telephone traffic engineering, a criterion that specifies the maximum number of calls or service demands that fail to receive immediate service. *Note:* The blocking criterion is usually expressed in probabilistic notation, such as P.001.

blocking factor. The number of records in a block. *Note:* The blocking factor is calculated by dividing the block length by the length of each record contained in the block. If the records are not of the same length, the average record length may be used to compute the blocking factor. *Synonym* grouping factor.

blocking formula. A specific probability distribution function intended to model calling patterns of users who fail to find available facilities. *Note:* There are several blocking formulas. The applicability of each to a given situation depends on its underlying assumptions regarding caller behavior.

blocking network. In telecommunications, a network that has fewer transmission paths than would be required if all users were to communicate simultaneously. *Note:* Blocking networks are used because not all users require service simultaneously. Certain statistical distributions apply to the patterns of user demand.

block length. The number of data units, such as bits, bytes, characters, or records, in a block.

block-loss probability. The ratio of the number of lost blocks to the total number of block transfer attempts during a specified period.

block-misdelivery probability. The ratio of the number of misdelivered blocks to the total number of block transfer attempts during a specified period.

block parity. The designation of one or more bits in a block as parity bits used to force the block into a selected parity, either odd or even. *Note:* Block parity is used to assist in error detection or correction.

block transfer. The process, initiated by a single action, of transferring one or more blocks of data.

block transfer attempt. A coordinated sequence of user and telecommunication system activities undertaken to effect transfer of an individual block from a source user to a destination user. *Note:* A block transfer attempt begins when the first bit of the block crosses the functional interface between the source user and the telecommunication system. A block transfer attempt ends either in successful block transfer or in block transfer failure.

block transfer efficiency. The average ratio of user information bits to total bits in successfully transferred blocks.

block transfer failure. Failure to deliver a block successfully. *Note:* The principal block transfer failure outcomes are: lost block, misdelivered block, and added block.

block transfer rate. The number of successful block transfers during a performance measurement period divided by the duration of the period.

block transfer time. The average value of the duration of a successful block transfer attempt. *Note:* A block transfer attempt is successful if (a) the transmitted block is delivered to the intended destination user within the maximum allowable performance period and (b) the contents of the delivered block are correct.

blue noise. In a spectrum of frequencies, a region in which the spectral density, *i.e.*, power per hertz, is proportional to the frequency.

blue-screening. *See* chroma keying.

blurring. In video systems, a global distortion, characterized by reduced sharpness of edges and limited spatial detail.

BOC. *Abbreviation for* Bell Operating Company.

Boltzmann's constant (k). The number that relates the average energy of a molecule to its absolute temperature. *Note:* Boltzmann's constant is approximately 1.38 x 10^{-23} J/K (joules/kelvin).

bond. An electrical connection that provides a low-resistance path between two conducting surfaces.

bonding. 1. In electrical engineering, the process of connecting together metal parts so that they make low resistance electrical contact for direct current and lower frequency alternating currents. 2. The process of establishing the required degree of electrical continuity between two or more conductive surfaces that are to be joined.

Boolean function. 1. A mathematical function that describes Boolean operations. 2. A switching function in which the number of possible values of the function and each of its independent variables is two.

Boolean operation. 1. Any operation in which each of the operands and the result take one of two values. *Note:* Typical states are "0 or 1," "on or off," "open or closed," or "present or absent." 2. An operation that follows the rules of Boolean Algebra.

bootstrap. 1. A technique or device designed to bring about a desired state by means of its own action. 2. That part of a computer program that may be used to establish another version of the computer program. 3. The automatic procedure whereby the basic operating system of a processor is reloaded following a complete shutdown or loss of memory. 4. A set of instructions that cause additional instructions to be loaded until the complete computer program is in storage. 5. To initialize a system by means of a bootstrap.

boresight. 1. The physical axis of a directional antenna. 2. To align a directional antenna, using either an optical procedure or a fixed target at a known location.

Bose-Chaudhuri-Hochquenghem code. *See* BCH code.

bound mode. In an optical fiber, a mode that (a) has a field intensity that decays monotonically in the transverse direction everywhere external to the core and (b) does not lose power to radiation. *Note:* Except for single-mode fibers, the power in bound modes is predominantly contained in the core of the fiber. *Synonyms* guided mode, trapped mode.

bound ray. *Synonym* guided ray.

bpi. *See* bits per inch.

braid. 1. An essential part of many fiber-optic cable designs, consisting of a layer of woven yarn. *Note:* In the case of single-fiber loose-buffered or two-fiber "zip-cord" loose-buffered fiber-optic cables, the braid is situated between the buffer tube and jacket. In the case of cables having multiple buffer tubes, the braid is usually situated between the inner jacket and outer jacket. 2. Loosely, an unwoven parallel bundle of yarn situated around the tight buffer of a single-fiber or two-fiber "zip-cord" fiber-optic cable. *Note 1:* The braid serves to add tensile strength to the cable. The braid may also be anchored to an optical connector or splice organizer assembly to secure the end of the cable. *Note 2:* The braid is often of an aramid yarn.

branch. 1. In a computer program, a conditional jump or departure from the implicit or declared order in which instructions are being executed. 2. To select a branch, as in definition #1. 3. A direct path joining two nodes of a network or graph. 4. In a power distribution system, a circuit from a distribution device (power panel) of a lower power handling capability than that of the input circuits to the device.

branching network. A network used for transmission or reception of signals over two or more channels.

branching repeater. A repeater with two or more outputs for each input. *Contrast with* multiport repeater.

breadboard. 1. An assembly of circuits or parts used to prove the feasibility of a device, circuit, system, or principle with little or no regard to the final configuration or packaging of the parts. 2. To prepare a breadboard.

break out. To separate the individual fibers or buffer tubes of a fiber-optic cable for the purpose of splicing or installing optical connectors. [After FAA] *Synonyms* fan out, furcate.

break-out box. A testing device that permits a user to access individual leads of an interface cable, using jumper wires, in order to monitor, switch, or patch the electrical output of the cable.

breakout cable. A multifiber fiber-optic cable design in which individual fibers, usually tight-buffered, are surrounded by separate strength members and jackets, which are in turn enveloped by a common jacket. *Note 1:* The breakout cable facilitates easy installation of fiber-optic connectors. All that need be done to prepare the ends of the cable to receive connectors is to remove the outer jacket, exposing what are essentially individual single-fiber cables. *Note 2:* Because it tends to induce bends in the fibers, the breakout cable design usually results in slightly higher transmission losses, for a given fiber, than loose-buffer designs. *Synonym* fanout cable.

breakout kit. A kit of materials, composed of an outer jacket in which is contained a strength member consisting of a bundle of usually aramid yarn, which jacket and yarn may be slipped over a loose buffer tube containing a single fiber, to convert the buffer tube and fiber to a complete single-fiber cable to which a fiber-optic connector may be directly attached. *Note 1:* A heat-shrinkable plastic boot may also be used for cosmetic purposes, strain relief, and to seal the point where the individual cables so created, merge. *Note 2:* Use of a breakout kit enables a fiber-optic cable containing multiple loose buffer tubes to receive connectors without the splicing of pigtails.

Brewster's angle. For a plane electromagnetic wavefront incident on a plane boundary between two dielectric media having different refractive indices, the angle of incidence at which transmittance from one medium to the other is unity when the wavefront is linearly polarized with its electric vector parallel to the plane of incidence. *Note 1:* Brewster=s angle θ_B, is given by

$$\theta_B = \tan^{-1}\left(\frac{n_2}{n_1}\right) = \tan^{-1}\sqrt{\frac{\varepsilon_2}{\varepsilon_1}} ,$$

where n_1 and n_2 are the refractive indices of the respective media, and ε_1 and ε_2, their respective electric permittivities. *Note 2:* For a randomly polarized ray incident at Brewster's angle, the reflected and refracted rays are at 90° with respect to one another.

Brewster's law. *See* Brewster's angle.

BRI. *Abbreviation for* basic rate interface.

brick. A colloquial name for a hand-held radiotelephone unit.

bridge. 1. In communications networks, a device that (a) links or routes signals from one ring or bus to another or from one network to another, (b) may extend the distance span and capacity of a single LAN system, (c) performs no modification to packets or messages, (d) operates at the data-link layer of the OSI–Reference Model (Layer 2), (e) reads packets, and (f) passes only those with addresses on the same segment of the network as the originating user. 2. A functional unit that interconnects two local area networks that use the same logical link control procedure, but may use different medium access control procedures. 3. A balanced electrical network, *e.g.*, a Wheatstone bridge. *Note:* A bridge may be used for electrical measurements, especially resistances or impedances. 4. *See* hybrid coil.

bridged ringing. The part of a signaling system in which ringers associated with a particular line are connected across that line.

bridge lifter. A device that electrically or physically removes bridged telephone pairs. *Note:* Relays, saturable inductors, and semiconductors are used as bridge lifters.

bridge-to-bridge station. A ship station operating in the port operations service in which messages are restricted to navigational communications and which is capable of operation from the ship's navigational bridge or, in the case of a dredge, from its main control station, operating on a frequency or frequencies in the 156-162 MHz band.

bridge transformer. *Synonym* hybrid coil.

bridging connection. A parallel connection used to extract some of the signal energy from a circuit, usually with negligible effect on the normal operation of the circuit.

bridging loss. At a given frequency, the loss that results when an impedance is connected across a transmission line. *Note:* Bridging loss is expressed as the ratio, in dB, of the signal power delivered, prior to bridging, to a given point in a system downstream from the bridging point, to the signal power delivered to the given point after bridging.

brightness. An attribute of visual perception in which a source appears to emit a given amount of light. *Note 1: "Brightness"* should be used only for nonquantitative references to physiological sensations and perceptions of light. *Note 2: "Brightness"* was formerly used as a synonym for the photometric term *"luminance"* and (incorrectly) for the radiometric term *"radiance."*

Brillouin diagram. *See* Brillouin scattering.

Brillouin scattering. In a physical medium, scattering of lightwaves, caused by thermally driven density fluctuations. *Note:* Brillouin scattering may cause frequency shifts of several gigahertz at room temperature.

broadband. *Synonym* wideband.

broadband exchange (BEX). A communications switch capable of interconnecting channels having bandwidths greater than voice bandwidth.

broadband ISDN (B-ISDN). An Integrated Services Digital Network (ISDN) offering broadband capabilities. *Note 1:* B-ISDN is a CCITT-proposed service that may (a) include interfaces operating at data rates from 150 to 600 Mb/s, (b) use asynchronous transfer mode (ATM) to carry all services over a single, integrated, high-speed packet-switched network, (c) have LAN interconnection capability, (d) provide access to a remote, shared disk server, (e) provide voice/video/data teleconferencing, (f) provide transport for programming services, such as cable TV, (g) provide single-user controlled access to remote video sources, (h) handle voice/video telephone calls, and (i) access shop-at-home and other information services. *Note 2:* Techniques used in the B-ISDN include code conversion, information compression, multipoint connections, and multiple-connection calls. Current proposals use a service-independent call structure that allows flexible arrangement and modular control of access and transport edges. The service components of a connection can provide each user with independent control of access features and can serve as the basis of a simplified control structure for multipoint and multiconnection calls. Such a network might be expected to offer a variety of ancillary information processing functions.

broadband system. *See* wideband.

broadcasting-satellite service. A radiocommunication service in which signals transmitted or retransmitted by space stations are intended for direct reception by the general public. In the broadcasting-satellite service, the term *"direct reception"* shall encompass both individual reception and community reception.

broadcasting satellite space station. A space station in the broadcasting-satellite service (sound broadcasting).

broadcasting service. A radiocommunication service in which the transmissions are intended for direct reception by the general public. This service may include sound transmissions, television transmissions or other types of transmissions.

broadcasting station. A station in the broadcasting service.

broadcast operation. The transmission of signals that may be simultaneously received by stations that usually make no acknowledgement.

broadside antenna. *Synonym* billboard antenna.

brouter. A combined bridge and router that operates without protocol restrictions, routes data using a protocol it supports, and bridges data it cannot route.

browser. Any computer software program for reading hypertext. *Note 1:* Browsers are usually associated with the Internet and the World Wide Web (WWW). *Note 2:* A browser may be able to access information in many formats, and through different services including HTTP, FTP, Gopher, and Archie.

browsing. [The] act of searching through automated information system storage to locate or acquire information without necessarily knowing of the existence or the format of the information being sought. [NIS]

b/s. *Abbreviation for* bits per second.

BSA. *Abbreviation for* basic serving arrangement.

BSE. *Abbreviation for* basic service element.

BSI. *Abbreviation for* British Standards Institution.

B6ZS. *Abbreviation for* bipolar with six-zero substitution. A T-carrier line code in which bipolar violations are deliberately inserted if user data contain a string of 6 or more consecutive zeros. *Note 1:* B6ZS is used to ensure a sufficient number of transitions to maintain system synchronization when the user data stream contains an insufficient number of "ones" to do so. *Note 2:* B6ZS is used in the North American hierarchy at the T2 rate.

B3ZS. *Abbreviation for* bipolar with three-zero substitution. A T-carrier line code in which bipolar violations are deliberately inserted if user data contain a string of 3 or more consecutive zeros. *Note 1:* B3ZS is used to ensure a sufficient number of transitions to maintain system synchronization when the user data stream contains an insufficient number of "ones" to do so. *Note 2:* B3ZS is used in the North American hierarchy at the T3 rate.

budgeting. *Synonym* proration (def. #1).

buffer. 1. A routine or storage medium used to compensate for a difference in rate of flow of data, or time of occurrence of events, when transferring data from one device to another. *Note:* Buffers are used for many purposes, such as (a) interconnecting two digital circuits operating at different rates, (b) holding data for use at a later time, (c) allowing timing corrections to be made on a data stream, (d) collecting binary data bits into groups that can then be operated on as a unit, (e) delaying the transit time of a signal in order to allow other operations to occur. 2. To use a buffer or buffers. 3. An isolating circuit, often an amplifier, used to minimize the influence of a driven circuit on the driving circuit. *Synonym* buffer amplifier. 4. In a fiber optic communication cable, one type of component used to encapsulate one or more optical fibers for the purpose of providing such functions as mechanical isolation, protection from physical damage and fiber identification. *Note:* The buffer may take the form of a miniature conduit, contained within the cable and called a loose buffer, or loose buffer tube, in which one or more fibers may be enclosed, often with a lubricating gel. A tight buffer consists of a polymer coating in intimate contact with the primary coating applied to the fiber during manufacture.

buffer amplifier. *Synonym* buffer (def. #3).

bug. 1. A concealed microphone or listening device or other audiosurveillance device. 2. A mistake in a computer program. 3. To install means for audiosurveillance. 4. A semiautomatic telegraph key. 5. A mistake or malfunction.

building out. The process of adding a combination of inductance, capacitance, and resistance to a cable pair so that its electrical length may be increased by a desired amount to control impedance and loss characteristics. *Synonym* line buildout.

bulk encryption. Simultaneous encryption of all channels of a multichannel telecommunications trunk. [NIS] *Note:* A single encryption device can be used to encrypt the output signal from a multiplexer.

bulletin board. A form of electronic messaging in which addressed messages or files are entered by users into a computer or network of computers. *Note:* Other users may obtain, at their convenience and request, messages or files available to them.

bunched frame-alignment signal. A frame-alignment signal in which the signal elements occupy consecutive digit positions.

bundle. A group of optical fibers or electrical conductors, such as wires and coaxial cables, usually in a single jacket. *Note:* Multiple bundles of optical fibers or electrical conductors may be placed in the same cable.

buried cable. *See* direct-buried cable.

burst. 1. In data communications, a sequence of signals, noise, or interference counted as a unit in accordance with some specific criterion or measure. 2. To separate continuous-form or multipart paper into discrete sheets.

burst isochronous. *Deprecated synonym for* isochronous burst transmission.

burst switching. In a packet-switched network, a switching capability in which each network switch extracts routing instructions from an incoming packet header to establish and maintain the appropriate switch connection for the duration of the packet, following which the connection is automatically released. *Note:* In concept, burst switching is similar to connectionless mode transmission, but it differs from the latter in that burst switching implies an intent to establish the switch connection in near real time so that only minimum buffering is required at the node switch.

burst transmission. 1. Transmission that combines a very high data signaling rate with very short transmission times. 2. Operation of a data network in which data transmission is interrupted at intervals. *Note:* Burst transmission enables communications between data terminal equipment (DTEs) and a data network operating at dissimilar data signaling rates. *Synonym* data burst.

bus. One or more conductors or optical fibers that serve as a common connection for a group of related devices.

bus interface unit (BIU). *See* network interface device.

bus network. *See* network topology.

bus topology. *See* network topology.

busy back. *Deprecated term. See* busy signal.

busy hour. In a communications system, the sliding 60-minute period during which occurs the maximum total traffic load in a given 24-hour period. *Note 1:* The busy hour is determined by fitting a horizontal line segment equivalent to one hour under the traffic load curve about the peak load point. *Note 2:* If the service time interval is less than 60 minutes, the busy hour is the 60-minute interval that contains the service timer interval. *Note 3:* In cases where more than one busy hour occurs in a 24-hour period, *i.e.,* when saturation occurs, the busy hour or hours most applicable to the particular situation are used. *Synonym* peak busy hour.**busy season.** During a 1-year cycle, the period of 3 consecutive months having the highest busy hour traffic.

busy signal. 1. In telephony, an audible or visual signal that indicates that no transmission path to the called number is available. *Synonym* busy tone. 2. In telephony, an audible or visual signal that indicates that the called number is occupied or otherwise unavailable. *Synonym* reorder tone.

busy test. In telephony, a test made to determine whether certain facilities, such as a subscriber line or a central office trunk, are available for use.

busy tone. *Synonym* busy signal.

busy verification. In a public switched telephone network, a network-provided service feature that permits an attendant to verify the busy or idle state of station lines and to break into the conversation. *Note:* A 440-Hz tone is applied to the line for 2 seconds, followed by a 0.5-second burst every 10 seconds, to alert both parties that the attendant is connected to the circuit.

BW. *Abbreviation for* bandwidth.

bypass. 1. The use of any telecommunications facilities or services that circumvents those of the local exchange common carrier. *Note:* Bypass facilities or services may be either customer-provided or vendor-supplied. 2. An alternate circuit that is routed around equipment or system component. *Note:* Bypasses are often used to allow system operation to continue when the bypassed equipment or a system component is inoperable or unavailable.

byte (B). A sequence of adjacent bits (usually 8) considered as a unit. *Note:* In pre-1970 literature, "*byte*" referred to a variable-length bit string. Since that time the usage has changed so that now it almost always refers to an 8-bit string. This usage predominates in computer and data transmission literature; when so used, the term is synonymous with "*octet.*"

cable. 1. An assembly of one or more insulated conductors, or optical fibers, or a combination of both, within an enveloping jacket. *Note:* A cable is constructed so that the conductors or fibers may be used singly or in groups. 2. A message sent by cable, or by any means of telegraphy.

cable assembly. A cable that is ready for installation in specific applications and usually terminated with connectors.

cable jacket. *See* sheath.

cable cutoff wavelength (ë$_{cc}$). For a cabled single-mode optical fiber under specified length, bend, and deployment conditions, the wavelength at which the fiber's second order mode is attenuated a measurable amount when compared to a multimode reference fiber or to a tightly bent single-mode fiber.

cable television relay service (CARS) station. A fixed or mobile station used for the transmission of television and related audio signals, signals of standard and FM broadcast stations, signals of instructional television fixed stations, and cablecasting from the point of reception to a terminal point from which the signals are distributed to the public.

cable TV (CATV). A television distribution method in which signals from distant stations are received, amplified, and then transmitted by (coaxial or fiber) cable or microwave links to users. *Note 1:* CATV originated in areas where good reception of direct broadcast TV was not possible. Now CATV also consists of a cable distribution system to large metropolitan areas in competition with direct broadcasting. *Note 2:* The abbreviation *CATV* originally meant *"community antenna television."* However, *CATV* is now usually understood to mean *cable TV.*

cache memory. A buffer, smaller and faster than main storage, used to hold a copy of instructions and data in main storage that are likely to be needed next by the processor and that have been obtained automatically from main storage.

call. 1. In communications, any demand to set up a connection. 2. A unit of traffic measurement. 3. The actions performed by a call originator. 4. The operations required to establish, maintain, and release a connection. 5. To use a connection between two stations. 6. The action of bringing a computer program, a routine, or a subroutine into effect, usually by specifying the entry conditions and the entry point.

call abandoned. *See* abandoned call.

call accepted signal. A call control signal sent by the called terminal to indicate that it accepts the incoming call.

call associated signaling (CAS). Signaling required for supervision of a bearer service between two end points, including support for the functions of call origination, call delivery, and handover.

call attempt. In a telecommunications system, a demand by a user for a connection to another user. *Note:* In telephone traffic analysis, call attempts are counted during a specific time frame. The call-attempt count includes all completed, overflowed, abandoned, and lost calls.

call back. [The] procedure for identifying a remote AIS terminal, whereby the host system disconnects the caller and then dials the authorized telephone number of the remote terminal to reestablish the connection.

call collision. 1. The contention that occurs when a terminal and data circuit-terminating equipment (DCE) specify the same channel at the same time to transfer a call request and handle an incoming call. *Note:* When call collision occurs, the DCE proceeds with the call request and cancels the incoming call. 2. The condition that occurs when a trunk or channel is seized at both ends simultaneously, thereby blocking a call. *Deprecated synonym* **glare.**

call completion rate. The ratio of successfully completed calls to the total number of attempted calls. *Note:* This ratio is typically expressed as either a percentage or a decimal fraction.

call control character. One of a set of control characters used in call-control signaling. *Note:* The signals representing call control characters may be used under defined conditions on interchange circuits other than the originating circuit.

call control signal. A member of the set of network management signals used to establish, maintain, or release a call.

call delay. 1. The delay that occurs when a call arrives at an automatic switching device and no channel or facility is immediately available to process the call. 2.

The time between the instant a system receives a call attempt and the instant of initiation of ringing at the call receiver end instrument.

call detail recording (CDR). A service feature in which call data on a specific telephone extension or group of subscribers are collected and recorded for cost-of-service accounting purposes.

call duration. 1. The time between (a) the instant a connection, *i.e.,* off-hook condition at each end, is established between the call originator and the call receiver and (b) the instant the call originator or the call receiver terminates the call. 2. In data transmission, the duration of the information transfer phase of an information transfer transaction.

called-line identification facility. A network-provided service feature in which the network notifies a calling terminal of the address to which the call has been connected.

called-line identification signal. A sequence of characters transmitted to the calling terminal to permit identification of the called line.

called party. *Synonym* call receiver.

called-party camp-on. A communication system service feature that enables the system to complete an access attempt in spite of issuance of a user blocking signal. *Note:* Systems that provide this feature monitor the busy user until the user blocking signal ends, and then proceed to complete the requested access. This feature permits holding an incoming call until the called party is free.

caller ID. A network service feature that permits the recipient of an incoming call to determine, even before answering, the number from which the incoming call is being placed.

caller identification. *See* caller ID.

call-failure signal. A signal sent in the backward direction indicating that a call cannot be completed because of a time-out, a fault, or a condition that does not correspond to any other particular signal.

call forwarding. A service feature, available in some switching systems, whereby calls can be rerouted automatically from one line, *i.e.,* station number, to another or to an attendant. *Note:* Call forwarding may be implemented in many forms.

call hold. A service feature in which a user may retain an existing call while accepting or originating another call using the same end instrument.

call identifier. A network utility that consists of a name assigned by the originating network for each established or partially established virtual call. *Note:* When a call identifier is used in conjunction with the calling data terminal equipment (DTE) address, the call identifier uniquely identifies the virtual call.

calling frequency. A radio frequency that a station uses to call another station.

calling-line identification facility. A network-provided service feature in which the network notifies a called terminal of the address from which the call has originated.

calling-line identification signal. A sequence of characters transmitted to the called terminal to permit identification of the calling line.

calling party. *Synonym* call originator.

calling-party camp-on. A service feature that enables the system to complete an access attempt in spite of temporary unavailability of system transmission or switching facilities required to establish the requested access. *Note:* Systems that provide calling party camp-on monitor the system facilities until the necessary facilities become available, and then proceed to complete the requested access. Such systems may or may not issue a system blocking signal to apprise the originating user of the access delay.

calling rate. The number of telephone calls originated during a specified time interval such as one hour.

calling sequence. A sequence of instructions together with any associated data necessary to perform a call.

calling signal. A call control signal transmitted over a circuit to indicate that a connection is desired.

call intensity. *Synonym* traffic intensity.

call management. 1. In telegraphy, route selection, signaling, and circuit usage and availability for a call. 2. In universal personal telecommunications, the ability of a user to inform the network how to handle incoming calls in accord with certain parameters, such as the call originator, the time of day, and the nature of the call. *Note:* Call management is accomplished by means of information in the user's service profile.

call-not-accepted signal. A call control signal sent by the called terminal to indicate that it does not accept the incoming call.

call originator. An entity, such as a person, equipment, or program that originates a call. *Synonym* calling party.

call pickup. A service feature of some switching systems enabling a user, by dialing a predetermined code, to answer incoming calls that are directed to another user in a preselected call group.

call processing. 1. The sequence of operations performed by a switching system from the acceptance of an incoming call through the final disposition of the call. 2. The end-to-end sequence of operations performed by a network from the instant a call attempt is initiated until the instant the call release is completed. 3. In data transmission, the operations required to complete all three phases of an information transfer transaction.

call progress signal. A call control signal transmitted by the called data circuit-terminating equipment (DCE) to the calling data terminal equipment (DTE) to report (a) the progress of a call by using a positive call progress signal or (b) the reason why a connection could not be established by using a negative call progress signal.

call progress tone. An audible signal returned by a network to a call originator to indicate the status of a call. *Note:* Examples of call progress tones include dial tones and busy signals.

call receiver. An entity, such as a person, equipment, or program to which a call is directed. *Synonym* called party.

call record. Recorded data pertaining to a single call.

call release time. In communication systems, the time interval from initiation of a clearing signal by a terminal until the available-line condition appears on originating terminal equipment.

call-request time. In the establishment of a connection or in the call setup, *i.e.*, placement of a call, the time from the initiation of a calling signal to the receipt of a proceed-to-select signal—such as a dial tone—by the call originator.

call restriction. A switching system service feature that prevents selected terminals from exercising one or more service features otherwise available from the switching system.

calls-barred facility. A service feature that permits a terminal either to make outgoing calls or to receive incoming calls, but not both.

call-second. A unit used to measure communications traffic. *Note 1:* A call-second is equivalent to 1 call 1 second long. *Note 2:* One user making two 75-second calls is equivalent to two users each making one 75-second call. Each case produces 150 call-seconds of traffic. *Note 3:* The CCS, equivalent to 100 call-seconds, is often used. *Note 4:* 3600 call-seconds = 36 CCS = 1 call-hour. *Note 5:* 3600 call-seconds per hour = 36 CCS per hour = 1 call-hour per hour = 1 erlang = 1 traffic unit.

call-selection time. In the establishment of a connection or the placement of a call, the time from the receipt by the call originator of a proceed-to-select signal (dial tone), until all the selection signals have been transmitted (dialing has been completed).

call set-up time. 1. The overall length of time required to establish a circuit-switched call between users. 2. For data communication, the overall length of time required to establish a circuit-switched call between terminals; *i.e.,* the time from the initiation of a call request to the beginning of the call message. *Note:* Call set-up time is the summation of: (a) call request time—the time from initiation of a calling signal to the delivery to the caller of a proceed-to-select signal; (b) selection time—the time from the delivery of the proceed-to-select signal until all the selection signals have been transmitted; and (c) post selection time—the time from the end of the transmission of the selection signals until the delivery of the call-connected signal to the originating terminal.

call sign. A station or address designator represented by a combination of characters or pronounceable words that is used to identify such entities as a communications facility, station, command, authority, activity, or unit.

call-sign allocation plan. The table of allocation of international call sign series contained in the current edition of the *International Telecommunication Union (ITU) Radio Regulations. Note:* In the table of allocation, the first two characters of each call sign (whether two letters or one number and one letter, in that order) identify the nationality of the station. In certain instances where the complete alphabetical block is allocated to a single nation, the first letter is sufficient for national identity. Individual assignments are made by appropriate national assignment authorities from the national allocation.

call spill-over. In common-channel signaling, the effect on a traffic circuit of the arrival at a switching center of an abnormally delayed call control signal relating to a previous call, while a subsequent call is being set up on the circuit.

call splitting. A switching system service feature that allows a switch attendant to talk privately in either direction on an established call.

call tracing. A procedure that permits an entitled user to be informed about the routing of data for an established connection, identifying the entire route from the origin to the destination. *Note:* There are two types of call tracing. Permanent call tracing permits tracing of all calls. On-demand call tracing permits tracing, upon request, of a specific call, provided that the called party dials a designated code immediately after the call to be traced is disconnected, *i.e.,* before another call is received or placed.

call transfer. A switching system service feature that allows the calling or called user to instruct the local switching equipment or switch attendant to transfer an existing call to another terminal. *Note:* Call transfer may be available on a call-by-call basis or on a semipermanent basis.

call waiting. In telephony, a service feature that provides an indication to a terminal already engaged in an established call that one or more calls are awaiting connection.

CAMA. *Acronym for* centralized automatic message accounting.

camp-on. *See* automatic callback, called-party camp-on, queue traffic.

camp-on busy signal. 1. A signal that informs a busy telephone user that another call originator is waiting for a connection. 2. A teleprinter exchange facility signal that automatically causes a calling station to retry the call-receiver number after a given interval when the call-receiver teleprinter is occupied or the circuits are busy. *Synonym* speed-up tone.

camp-on-with-recall. A camp-on with the release of the call-originator terminal until the called-receiver terminal becomes free. *Note:* The call originator can thus establish other calls until the recall signal is obtained, rather than simply wait until the call-receiver line is available.

CAN. *Abbreviation for* cancel character.

Canadian Standards Association (CSA). An independent, nongovernment, not-for-profit association for the development, by consensus, of Canadian standards and product certification.

cancel character (CAN). 1. A control character used by some conventions to indicate that the data with which it is associated are in error or are to be disregarded. 2. An accuracy control character used to indicate that the data with which it is associated are in error, are to be disregarded, or cannot be represented on a particular device.

capacitive coupling. The transfer of energy from one circuit to another by means of the mutual capacitance between the circuits. *Note 1:* The coupling may be deliberate or inadvertent. *Note 2:* Capacitive coupling favors transfer of the higher frequency components of a signal, whereas inductive coupling favors lower frequency components, and conductive coupling favors neither higher nor lower frequency components.

capacity. *See* channel capacity, traffic capacity.

capture effect. A phenomenon, associated with FM reception, in which only the stronger of two signals at or near the same frequency will be demodulated. *Note 1:* The complete suppression of the weaker signal occurs at the receiver limiter, where it is treated as noise and rejected. *Note 2:* When both signals are nearly equal in strength, or are fading independently, the receiver may switch from one to the other. *Synonym* FM capture effect.

cardinal radial. Those eight radials at 0°, 45°, 90°, 135°, 180°, 225°, 270°, and 315° of azimuth with respect to true north. *Note:* The four radials at 0°, 90°, 180°, and 270° of azimuth with respect to true north are referred to as the cardinal points. The cardinal points are equivalent to true north, east, south, and west.

carrier. *Synonym* common carrier.

carrier (cxr). 1. A wave suitable for modulation by an information-bearing signal. 2. An unmodulated emission. *Note:* The carrier is usually a sinusoidal wave or a uniform or predictable series of pulses. *Synonym* carrier wave. 3. Sometimes employed as a *synonym for* carrier system.

carrier dropout. A short-duration loss of carrier signal.

carrier frequency. 1. The nominal frequency of a carrier wave. 2. In frequency modulation, *synonym* center frequency.

carrier leak. The carrier remaining after carrier suppression in a suppressed carrier transmission system. *Note:* Sometimes the residual carrier is used to provide the reference for an automatic frequency control system.

carrier level. The level of a carrier signal at a specified point in a communications system. *Note:* The carrier level is usually expressed in dB relative to a specified reference level.

carrier multiplex. *See* frequency-division multiplexing.

carrier noise level. The noise level resulting from undesired variations of a carrier in the absence of any intended modulation. *Synonym* residual modulation.

carrier power (of a radio transmitter). 1. The average power supplied to the antenna transmission line by a transmitter during one radio frequency cycle taken under the condition of no modulation. *Note:* The concept does not apply to pulse modulation or frequency-shift keying. 2. The average unmodulated power supplied to a transmission line.

carrier sense. In a local area network, an ongoing activity of a data station to detect whether another station is transmitting.

carrier sense multiple access (CSMA). A network control scheme in which a node verifies the absence of other traffic before transmitting.

carrier sense multiple access with collision avoidance (CSMA/CA). A network control protocol in which (a) a carrier sensing scheme is used, (b) a data station that intends to transmit sends a jam signal, (c) after waiting a sufficient time for all stations to receive the jam signal, the data station transmits a frame, and (d) while transmitting, if the data station detects a jam signal from another station, it stops transmitting for a random time and then tries again.

carrier sense multiple access with collision detection (CSMA/CD). A network control protocol in which (a) a carrier sensing scheme is used and (b) a transmitting data station that detects another signal while transmitting a frame, stops transmitting that frame, transmits a jam signal, and then waits for a random time interval before trying to send that frame again.

carrier shift. 1. In the transmission of binary or teletypewriter signals, keying in which the carrier frequency is shifted in one direction for marking signals and in the opposite direction for spacing signals. 2. In amplitude modulation, a condition that results from imperfect modulation in which the positive and negative excursions of the modulating envelope are unequal in amplitude. *Note 1:* The carrier shift results in a change in carrier power. *Note 2:* The carrier shift may be a shift to a higher or to a lower frequency.

carrier suppression. *See* suppressed carrier transmission.

carrier synchronization. In a radio receiver, the generation of a reference carrier with a phase closely matching that of a received signal.

carrier system. A multichannel telecommunications system in which a number of individual circuits (data, voice, or combination thereof) are multiplexed for transmission between nodes of a network. *Note 1:* In carrier systems, many different forms of multiplexing may be used, such as time-division multiplexing and frequency-division multiplexing. *Note 2:* Multiple layers of multiplexing may ultimately be performed upon a given input signal; *i.e.,* the output resulting from one stage of modulation may in turn be modulated. *Note 3:* At a given node, specified channels, groups, supergroups, *etc.,* may be demultiplexed without demultiplexing the others. *Synonym (loosely)* carrier.

carrier-to-noise ratio (CNR). In radio receivers, the ratio of the level of the carrier to that of the noise in the intermediate frequency (IF) band before any nonlinear process, such as amplitude limitation and detection, takes place. *Note:* The CNR is usually expressed in dB.

carrier-to-receiver noise density (*C*/k*T*). In satellite communications, the ratio of the received carrier power to the receiver noise power density. *Note 1:* The carrier-to-receiver noise density ratio is usually expressed in dB. *Note 2:* The carrier-to-receiver noise density is given by $C/\mathrm{k}T$, where C is the received carrier power in watts, k is Boltzmann's constant in joules per kelvin, and T is the receiver system noise temperature in kelvins. *Note 3:* The receiver noise power density, kT, is the receiver noise power per hertz.

carrier wave. *Synonym* carrier (cxr) (def. #2).

CARS. *Acronym for* cable television relay service.

Carson bandwidth rule. A rule defining the approximate bandwidth requirements of communications system components for a carrier signal that is frequency modulated by a continuous or broad spectrum of frequencies rather than a single frequency. *Note 1:* The Carson bandwidth rule is expressed by the relation CBR = $2(\Delta f + f_m)$ where CBR is the bandwidth requirement, Δf is the carrier peak deviation frequency, and f_m is the highest modulating frequency. *Note 2:* The Carson bandwidth rule is often applied to transmitters, antennas, optical sources, receivers, photodetectors, and other communications system components.

CAS. *Abbreviation for* centralized attendant services.

CASE. *Acronym for* computer-aided software engineering, computer-aided systems engineering. Software used for the automated development of systems software, *i.e.*, computer code. *Note 1:* CASE functions include analysis, design, and programming. *Note 2:* CASE tools automate methods for designing, documenting, and producing structured computer code in the desired programming language.

case shift. 1. In data equipment, the change from letters to other characters, or vice versa. 2. In typewriting or typesetting, the change from lower case letters to upper case letters, or vice versa.

CASE technology. Technology that makes use of computer assisted software engineering (CASE) to enhance the development of systems design and development.

Cassegrain antenna. An antenna in which the feed radiator is mounted at or near the surface of a concave main reflector and is aimed at a convex secondary reflector slightly inside the focus of the main reflector. *Note 1:* Energy from the feed unit illuminates the secondary reflector, which reflects it back to the main reflector, which then forms the desired forward beam. *Note 2:* The Cassegrain antenna design is adapted from optical telescope technology and allows the feed radiator to be more easily supported.

catastrophic degradation. The rapid reduction of the ability of a system, subsystem, component, equipment, or software to perform its intended function. *Note:* Catastrophic degradation usually results in total failure to perform any function.

Category 3. The ANSI/EIA/TIA-568 designation for 100-ohm unshielded twisted-pair cables and associated connecting hardware whose characteristics are specified for data transmission up to 16 Mb/s.

Category 4. The ANSI/EIA/TIA-568 designation for 100-ohm unshielded twisted-pair cables and associated connecting hardware whose characteristics are specified for data transmission up to 20 Mb/s.

Category 5. The ANSI/EIA/TIA-568 designation for 100-ohm unshielded twisted-pair cables and associated connecting hardware whose characteristics are specified for data transmission up to 100 Mb/s.

CATV. *Abbreviation for* cable TV.

cavity. A volume defined by conductor-dielectric or dielectric-dielectric reflective boundaries, or a combination of both, and having dimensions designed to produce specific interference effects (constructive or destructive) when excited by an electromagnetic wave.

C-band. *Colloquially,* a frequency band between 4 GHz and 6 GHz used in satellite communications. *Note 1:* For procurement purposes, the radio frequency band(s) must be specified using the upper and lower limits of the band, per 47 CFR 300. *Note 2:* Letter designators of radio frequency bands are imprecise, deprecated, and legally obsolete.

CCH. *Abbreviation for* connections per circuit hour.

CCIR. *Abbreviation for* International Radio Consultative Committee; a predecessor organization of the ITU-T.

CCIS. *Abbreviation for* common-channel interoffice signaling.

CCITT. *Abbreviation for* International Telegraph and Telephone Consultative Committee; a predecessor organization of the ITU-T.

CCS. *Abbreviation for* hundred call-seconds. *See* call-second.

CCSA. *Abbreviation for* common control switching arrangement.

CDF. *Abbreviation for* combined distribution frame.

CDMA. *Abbreviation for* code-division multiple access.

CDPSK. *Abbreviation for* coherent differential phase-shift keying.

CDR. *Abbreviation for* call detail recording.

CD ROM. *Abbreviation for* compact disk read-only memory. An optical digital storage device, of high capacity, capable of being read from but not written to.

C-E. *Abbreviation for* communications-electronics.

CEI. *Abbreviation for* comparably efficient interconnection.

cell. 1. In cellular mobile, the geographical area covered by the smaller of: a base station, or a subsystem (sector antenna) of that base station corresponding to a specific logical identification on the radio path. *Note:* Mobile stations in a cell may be reached by the corresponding radio equipment of the base station. 2. In communications, a string that contains a header and user information. *Note 1:* A cell is dedicated to one user for one session. Cells for a given system are usually of fixed length and smaller than a frame, such as 424 bits for a cell, compared to 1024 for a frame. *Note 2:* In asynchronous transfer mode (ATM) systems, a cell consists of 53 bytes, *i.e.,* a 5-byte header field and a 48-byte information field. *Note 3:* A cell does not have error-correction capability and is therefore suited for low-BER communications systems, such as digital fiber optic systems. 3. In OSI, a fixed-length block labeled at the Physical Layer of the Open Systems Interconnection—Reference Model (OSI—RM). 4. In computer systems, an addressable, internal hardware location. 5. In computer applications, a single location on a spreadsheet.

cell relay. A statistically multiplexed interface protocol for packet switched data communications that uses fixed-length packets, *i.e.,* cells, to transport data. *Note 1:* Cell relay transmission rates usually are between 56 kb/s and 1.544 Mb/s, *i.e.,* the data rate of a DS1 signal. *Note 2:* Cell relay protocols (a) have neither flow control nor error correction capability, (b) are information-content independent, and (c) correspond only to layers one and two of the ISO Open Systems Interconnection—Reference Model. *Note 3:* Cell relay systems enclose variable-length user packets in fixed-length packets, *i.e.,* cells, that add addressing and verification information. Frame length is fixed in hardware, based on time delay and user packet-length considerations. One user data message may be segmented over many cells. *Note 4:* Cell relay is an implementation of fast packet technology that is used in (a) connection-oriented broadband integrated services digital networks (B-ISDN) and (b) connectionless IEEE 802.6, switched multi-megabit data service (SMDS). *Note 5:* Cell relay is used for time-sensitive traffic such as voice and video.

cellular mobile. A mobile communications system that uses a combination of radio transmission and conventional telephone switching to permit telephone communication to and from mobile users within a specified area. *Note:* In cellular mobile systems, large geographical areas are segmented into many smaller areas, *i.e.*, cells, each of which has its own radio transmitters and receivers and a single controller interconnected with the public switched telephone network. *Synonyms* cellular phone, cellular radio, cellular telephone.

cellular phone. *Synonym* cellular mobile.

cellular radio. *Synonym* cellular mobile.

cellular telephone. *Synonym* cellular mobile.

CELP. *Acronym for* code-excited linear prediction.

center frequency. 1. In frequency modulation, the rest frequency, *i.e.*, the frequency of the unmodulated carrier. *Synonym* carrier frequency. 2. In facsimile systems, the frequency midway between the picture-black and picture-white frequencies.

centralized attendant services (CAS). A function of a usually centrally located attendant console that permits the control of multiple switches, some of which may be geographically remote.

centralized automatic message accounting (CAMA). An automatic message accounting system that serves more than one switch from a central location. *Note:* When using CAMA, human intervention may be required.

centralized operation. Operation of a communication network in which transmission may occur between the control station and any tributary station, but not between tributary stations.

centralized ordering group (COG). An organization provided by some communications service providers to coordinate services between the companies and vendors.

central office (C.O.). A common carrier switching center in which trunks and loops are terminated and switched. *Note:* In the DOD, "common carrier" is called *"commercial carrier." Synonyms* exchange, local central office, local exchange, local office, switching center (except in DOD DSN [formerly AUTOVON] usage), switching exchange, telephone exchange. *Deprecated synonym* switch.

central office connecting facility. *Loosely*, in the sense of a trunk between public and private switches, *a synonym for* **central office trunk.** *See* trunk.

central office trunk. *Loosely*, in the sense of a trunk between public and private switches, *a synonym for* **central office connecting facility.** *See* trunk.

central processing unit. *See* CPU.

central processor. *Synonym* CPU.

Centrex® (CTX) service. A service offered by Bell Operating Companies that provides functions and features comparable to those provided by a PBX or a PABX. *Note: "Centrex® C.O."* indicates that all equipment except the attendant's position and station equipment is located in the central office. *"Centrex® C.U."* indicates that all equipment, including the dial switching equipment, is located on the customer's premises.

certification. [The] comprehensive evaluation of the technical and nontechnical security features of an AIS [automated information system] and other safeguards, made in support of the accreditation process, to establish the extent to which a particular design and implementation meets a set of specified security requirements.

certified network engineer. In computer networking, one who has met proprietary training and certification requirements pertinent to network design or maintenance. *Note:* The training requirements embrace both software and hardware configuration.

cesium clock. A clock containing a cesium standard as a frequency-determining element.

cesium standard. A primary frequency standard in which electronic transitions between the two hyperfine ground states of cesium-133 atoms is used to control the output frequency. *Note*: The energy level between the two hyperfine ground states corresponds, in the absence of external influences (*e.g.*, the magnetic field of the Earth), to a frequency of 9,192,631,770 Hz.

chad. The material separated from a punched tape or a punched card when forming a hole.

chadless tape. 1. Punched tape that has been punched in such a way that chad is not formed. 2. A punched tape in which only partial perforation is performed so that the chad remains attached to the tape. *Note:* The partial perforation is deliberate, and should not be confused with imperfect chadding.

chad tape. Punched tape used in telegraphy/teletypewriter operation.

channel. 1. A connection between initiating and terminating nodes of a circuit. 2. A single path provided by a transmission medium via either (a) physical separation, such as by multipair cable or (b) electrical separation, such as by frequency- or time-division multiplexing. 3. A path for conveying electrical or electromagnetic signals, usually distinguished from other parallel paths. 4. Used in conjunction with a predetermined letter, number, or codeword to reference a specific radio frequency. 5. The portion of a storage medium, such as a track or a band, that is accessible to a given reading or writing station or head. 6. In a communications system, the part that connects a data source to a data sink.

channel-associated signaling. Signaling in which the signals necessary to switch a given circuit are transmitted via the circuit itself or via a signaling channel permanently associated with it.

channel bank. The part of a carrier-multiplex terminal that performs the first step of modulation by multiplexing a group of channels into a higher bandwidth analog channel or higher bit-rate digital channel and, conversely, demultiplexes these aggregates back into individual channels.

channel bits. Binary data transmitted over a communications link. *Note:* Channel bits are derived from user information by FEC (forward error correction) coding and interleaving.

channel capacity. The maximum possible information transfer rate through a channel, subject to specified constraints.

channel gate. A device for connecting a channel to a highway, or a highway to a channel, at specified times.

channelization. The use of a single wideband, *i.e.,* high-capacity, facility to create many relatively narrowband, *i.e.,* lower capacity, channels by subdividing the wideband facility.

channel noise level. 1. The ratio of the channel noise at any point in a transmission system to an arbitrary level chosen as a reference. *Note 1:* The channel noise level may be expressed in (a) dB above reference noise (dBrn), (b) dB above reference noise with C-message weighting (dBrnC), or (c) adjusted dB (dBa). *Note 2:* Each unit used to measure channel noise level reflects a circuit noise reading of a specialized instrument designed to account for different interference effects that occur under specified conditions. 2. The noise power density spectrum in the frequency range of interest. 3. The average noise power in the frequency range of interest.

channel offset. The constant frequency difference between a channel frequency and a reference frequency which may frequency hop.

channel packing. Maximizing the use of voice frequency channels for data transmission by multiplexing a number of channels of lower data rate into a single voice frequency channel of higher data rate.

channel reliability (ChR). The percentage of time a channel was available for use in a specified period of scheduled availability. *Note 1:* Channel reliability is given by

$$\text{ChR} = 100\left(1 - \frac{T_0}{T_s}\right) = 100\,\frac{T_a}{T_s},$$

where T_o is the channel total outage time, T_s is the channel total scheduled time, and T_a is the channel total available time. *Note 2:* $T_s = T_a + T_o$.

channel service unit (CSU). A line bridging device that (a) is used to perform loop-back testing, (b) may perform bit stuffing, (c) may also provide a framing and formatting pattern compatible with the network, and (d) is the last signal regeneration point, on the loop side, coming from the central office; before the regenerated signal reaches a multiplexer or data terminal equipment (DTE).

channel supergroup. *See* group.

channel time slot. A time slot that starts at a particular instant in a frame and is allocated to a channel for transmitting data, such as a character or an in-slot signal.

character. 1. A letter, digit, or other symbol that is used as part of the organization, control, or representation of data. 2. One of the units of an alphabet.

character check. A method of error detection using the preset rules for the formulation of characters.

character-count integrity. The preservation of the exact number of characters that are originated in a message in the case of message communications, or per unit time, in the case of a user-to-user connection. *Note:* Character-count integrity is not the same as character integrity, which requires that the characters delivered are, in fact, exactly the same as they were originated.

character filter. Software that is capable of selectively removing characters from a data stream, *e.g.*, software that removes communications-control characters so that they are not printed.

character generator. A functional unit that converts the coded representation of a character into the graphic representation of the character for display.

character integrity. Preservation of a character during processing, storage, and transmission.

character interval. In a communications system, the total number of unit intervals required to transmit any given character, including synchronizing, information, error checking, or control characters, but not including signals that are not associated with individual characters. *Note:* An example of a time interval that is excluded when determining character interval is any time added between the end of a stop signal and the beginning of the next start signal to accommodate changing transmission conditions, such as a change in data signaling rate or buffering requirements. This added time is defined as a part of the intercharacter interval.

characteristic distortion. In telegraphy, the distortion caused by transients that, as a result of previous modulation, are present in the transmission channel. *Note:* Characteristic distortion effects are not consistent. Their effects on a given signal transition are dependent upon transients remaining from previous signal transitions.

characteristic frequency. A frequency which can be easily identified and measured in a given emission. A carrier frequency may, for example, be designated as the characteristic frequency.

characteristic impedance (Z_0). 1. The impedance of a circuit that, when connected to the output terminals of a uniform transmission line of arbitrary length, causes the line to appear infinitely long. *Note 1:* A uniform line terminated in its characteristic impedance will have no standing waves, no reflections from the end, and a constant ratio of voltage to current at a given frequency at every point on the line. *Note 2:* If the line is not uniform, the iterative impedance must be used. 2. For Maxwell's equations, the impedance of a linear, homogeneous, isotropic, dielectric propagation medium free of electric charge, given by the relation $Z=(\mu/\varepsilon)^2$ where μ is the magnetic permeability and ε is the electric permittivity of the medium.

character recognition. The identification of characters by automatic means.

character set. 1. A finite set of different characters that is complete for a given purpose. *Note:* A character set may or may not include punctuation marks or other symbols. 2. An ordered set of unique representations. *Note:* Examples of character sets include the 26 letters of the English alphabet, Boolean characters 0 and 1, the 128 ASCII characters, and International Telegraph Alphabet 5 (ITA-5), published as CCITT Recommendation V.3 and ISO 646.

characters per inch (cpi). In a recording medium, a unit of linear packing density of characters.

characters per second (cps). A unit of signaling speed used to express the number of characters passing a designated point per second.

character-stepped. Pertaining to control of start-stop teletypewriter equipment in which a device is stepped one character at a time. *Note:* The step interval is equal to or greater than the character interval at the applicable signaling rate.

charging reference location. In Universal Personal Telecommunications Service, the geographical location that may be used by the UPT service providers to determine the distance-related charges applying to the call originator and/or to the destination UPT user.

check. A process for determining accuracy.

check bit. A bit, such as a parity bit, derived from and appended to a bit string for later use in error detection and possibly error correction.

check character. A character, derived from and appended to a data item, for later use in error detection and possibly error correction.

check digit. A digit, derived from and appended to a data item, for later use in error detection and possibly error correction.

checksum. 1. The sum of a group of data items, which sum is used for checking purposes. *Note 1:* A checksum is stored or transmitted with the group of data items. *Note 2:* The checksum is calculated by treating the data items as numeric values. *Note 3:* Checksums are used in error detecting and correcting. 2. [The] value computed, via some parity or hashing algorithm, on information requiring protection against error or manipulation. *Note:* Checksums are stored or transmitted with data and are intended to detect data integrity problems.

chip. 1. *Synonym* integrated circuit. 2. In satellite communications systems, the smallest element of data in an encoded signal. 3. The most elemental component of a spread spectrum signal when it is decompressed in time; that is, the longest duration signal in which signal parameters are approximately constant.. 4. In micrographic and display systems, a relatively small and separate piece of microform that contains microimages and coded information for search, identification, and retrieval purposes.

chip rate. 1. The rate of encoding. 2. In direct-sequence-modulation spread-spectrum systems, the rate at which the information signal bits are transmitted as a pseudorandom sequence of chips. *Note:* The chip rate is usually several times the information bit rate.

chip time. In spread-spectrum systems, the duration of a chip produced by a frequency-hopping signal generator.

chirping. 1. The rapid changing, as opposed to long-term drifting, of the frequency of an electromagnetic wave. *Note:* Chirping is most often observed in pulsed operation of a source. 2. A pulse compression technique that uses (usually linear) frequency modulation during the pulse.

chroma keying. In television, nearly instantaneous switching between multiple video signals, based on the state, *i.e.*, phase, of the color (chroma) signal of one, to form a single composite video signal. *Note 1:* Chroma keying is used to create an overlay effect in the final picture, *e.g.*, to insert a false background, such as a weather map or scenic view, behind the principal subject being photographed. *Note 2:* The principal subject is photographed against a background having a single color or a relatively narrow range of colors, usually in the blue. When the phase of the chroma signal corresponds to the preprogrammed state or states associated with the background color, or range of colors, behind the principal subject, the signal from the alternate, *i.e.*, false, background is inserted in the composite signal and presented at the output. When the phase of the chroma signal deviates from that associated with the background color(s) behind the principal subject, video associated with the principal subject is presented at the output. *Synonym* color keying. *Colloquial synonym* blue-screening. *Contrast with* chrominance signal, composite video.

chromatic dispersion. *A commonly used (but redundant) synonym for* material dispersion. *See* dispersion.

chrominance signal. In color television, that signal or portion of the composite signal that bears the color information.

cipher. 1. [A] cryptographic system in which units of plain text are substituted according to a predetermined key. *Note:* A cipher is any cryptographic system in which arbitrary symbols, or groups of symbols, represent units of plain text of regular length, usually single letters, or in which units of plain text are rearranged, or both, in accordance with certain predetermined rules. 2. The result of using a cipher. *Note:* An example of a cipher is an enciphered message or text.

cipher system. Any cryptosystem that requires the use of a key to convert, unit by unit, plain text, encoded text, or signals into an unintelligible form for secure transmission. *Note:* The capability to decipher must be available at the receiving site.

cipher text. Enciphered information. *Note:* Cipher text is the result obtained from enciphering plain or encoded text.

ciphony. The process of enciphering audio information. *Note:* "*Ciphony*" is a contraction of "*ciphered telephony.*"

circuit. 1. The complete path between two terminals over which one-way or two-way communications may be provided. 2. An electronic path between two or more points, capable of providing a number of channels. 3. A number of conductors connected together for the purpose of carrying an electrical current. 4. An electronic closed-loop path among two or more points used for signal transfer. 5. A number of electrical components, such as resistors, inductances, capacitors, transistors, and power sources connected together in one or more closed loops.

circuit noise level. At any point in a transmission system, the ratio of the circuit noise at that point to an arbitrary level chosen as a reference. *Note:* The circuit noise level is usually expressed in dBrn0, signifying the reading of a circuit noise meter, or in dBa0, signifying circuit noise meter reading adjusted to represent an interfering effect under specified conditions.

circuit reliability (*CiR*). The percentage of time a circuit was available for use in a specified period of scheduled availability. *Note 1:* Circuit reliability is given by

$$\mathrm{CiR} = 100\left(1 - \frac{T_0}{T_s}\right) = 100\ \frac{T_a}{T_s},$$

where T_o is the circuit total outage time, T_s is the circuit total scheduled time, and T_a is the circuit total available time. *Note 2:* $T_s = T_a + T_o$. *Synonym* time availability.

circuit restoration. The process by which a communications circuit is established between two users after disruption or loss of the original circuit. *Note:* Circuit restoration is usually performed in accordance with planned procedures and priorities. Restoration may be effected automatically, *e.g.*, by switching to a hot standby, or manually, *e.g.*, by manual patching.

circuit routing: In open systems architecture, the logical path of a message in a communications network based on a series of gates at the physical network layer in the Open Systems Interconnection–Reference Model and the GOSIP FIPS PUB 146-1.

circuit-switched data transmission service. A data transmission service requiring the establishment of a circuit-switched connection before data can be transferred from source data terminal equipment (DTE) to a sink DTE. *Note:* A circuit-switched data transmission service uses a connection-oriented network.

circuit switching. 1. A method of routing traffic through a switching center, from local users or from other switching centers, whereby a connection is established between the calling and called stations until the connection is released by the called or calling station. 2. A process that, on demand, connects two or more data terminal equipments (DTEs) and permits the exclusive use of a data circuit between them until the connection is released.

circuit switching center. *See* **circuit switching, switching center.**

circuit switching unit (CSU). Equipment used for routing messages over common-user circuits that interconnect a source data terminal equipment (DTE) to a sink DTE for information interchange.

circuit transfer mode. In ISDN applications, a transfer mode by means of permanent allocation of channels or bandwidth between connections.

circular polarization. In electromagnetic wave propagation, polarization such that the tip of the electric field vector describes a helix. *Note 1:* The magnitude of the electric field vector is constant. *Note 2:* The projection of the tip of the electric field vector upon any fixed plane intersecting, and normal to, the direction of propagation, describes a circle. *Note 3:* A circularly polarized wave may be resolved into two linearly polarized waves in phase quadrature with their planes of polarization at right angles to each other. *Note 4:* Circular polarization may be referred to as *"right-hand"* or *"left-hand,"* depending on whether the helix describes the thread of a right-hand or left-hand screw, respectively.

circulator. 1. A passive junction of three or more ports in which the ports can be accessed in such an order that when power is fed into any port it is transferred to the next port, the first port being counted as following the last in order. 2. In radar, a device that switches the antenna alternately between the transmitter and receiver.

civision. 1. The application of cryptography to television signals. 2. Television signals that have been enciphered to preserve the confidentiality of the transmitted information.

C/kT. *Abbreviation for* carrier-to-receiver noise density.

cladding. 1. Of an optical fiber, one or more layers of material of lower refractive index, in intimate contact with a core material of higher refractive index. 2. A process of covering one metal with another (usually achieved by pressure rolling, extruding, drawing, or swaging) until a bond is achieved.

cladding diameter. In the cross section of a realizable optical fiber, ideally circular, but in practice assumed to a first approximation to be elliptical, the average of the diameters of the smallest circle that can be circumscribed about the cladding, and the largest circle that can be inscribed within the cladding.

cladding eccentricity. *See* ovality.

cladding mode. An undesired mode that is confined to the cladding of an optical fiber by virtue of the fact that the cladding has a higher refractive index than the surrounding medium, *i.e.,* air or primary polymer overcoat. *Note:* Modern fibers have a primary polymer overcoat with a refractive index that is slightly higher, rather than lower, than that of the cladding, in order to strip off cladding modes after only a few centimeters of propagation.

cladding mode stripper. A device for converting optical fiber cladding modes to radiation modes; as a result, the cladding modes are removed from the fiber. *Note:* Often a material such as the fiber coating or jacket having a refractive index equal to or greater than that of the fiber cladding will perform this function.

cladding noncircularity. *See* ovality.

cladding ovality. *Synonym* cladding noncircularity. *See* ovality.

cladding ray. *See* cladding mode.

C-language. A general-purpose, high-level, structured computer programming language. *Note:* C-language was originally designed for and implemented on the UNIX™ operating system.

CLASS. *Acronym for* custom local area signaling services.

class d address. *Synonym (in Internet protocol)* multicast address.

classmark. A designator used to describe the service feature privileges, restrictions, and circuit characteristics for lines or trunks that access a switch. *Note:* Examples of classmarks include precedence level, conference privilege, security level, and zone restriction. *Synonym* class-of-service mark.

class of emission. The set of characteristics of an emission, designated by standard symbols, *e.g.*, type of modulation of the main carrier, modulating signal, type of information to be transmitted, and also, if appropriate, any additional signal characteristics.

class of office. A ranking, assigned to each switching center in a communications network, determined by the center switching functions, interrelationships with other offices, and transmission requirements.

class of service. 1. A designation assigned to describe the service treatment and privileges given to a particular terminal. 2. A subgrouping of telephone users for the purpose of rate distinction. *Note:* Examples of class of service subgrouping include distinguishing between (a) individual and party lines, (b) Government and non-Government lines, (c) those permitted to make unrestricted international dialed calls and those not so permitted, (d) business, residence, and coin-operated, (e) flat rate and message rate, and (f) restricted and extended area service. 3. A category of data transmission provided by a public data network in which the data signaling rate, the terminal operating mode, and the code structure, are standardized. *Note:* Class of service is defined in CCITT Recommendation X.1. *Synonym* user service class.

class-of-service mark. *Synonym* classmark

clear. To cause one or more storage locations to be in a prescribed state, usually that corresponding to a zero or that corresponding to the space character.

clear channel. 1. In radio broadcasting, a frequency assigned for the exclusive use of one entity. 2. In networking, a signal path that provides its full bandwidth for a user's service. *Note:* No control or signaling is performed on this path.

clear collision. Contention that occurs when a DTE and a DCE simultaneously transfer a clear request packet and a clear indication packet specifying the same logical channel. *Note:* The DCE will consider that the clearing is completed and will not transfer a DCE clear confirmation packet.

clear confirmation signal. A call control signal used to acknowledge reception of the data-terminal-equipment (DTE) clear request by the data circuit-terminating equipment (DCE) or to acknowledge the reception of the DCE clear indication by the DTE.

clearing. 1. A sequence of events used to disconnect a call and return to the ready state. 2. Removal of data from an AIS, its storage devices, and other peripheral devices with storage capacity, in such a way that the data may not be reconstructed using normal system capabilities (*i.e.*, through the keyboard). *Note:* An AIS need not be disconnected from any external network before clearing takes place. Clearing enables a product to be reused within, but not outside of, a secure facility. It does not produce a declassified product by itself, but may be the first step in the declassification process.

clear message. 1. A message that (a) is sent in the forward direction and the backward direction, (b) contains a circuit-released signal or circuit-released acknowledgment signal, and (c) usually contains an indication of whether the message is in the forward or the backward direction. 2. A message in plain language, *i.e.,* not enciphered.

clear text. *Synonym* plain text.

cleave. 1. In an optical fiber, a deliberate, controlled break, intended to create a perfectly flat endface, perpendicular to the longitudinal axis of the fiber. *Note:* A cleave is made by first introducing a microscopic fracture ("nick") into the fiber with a special tool, called a "*cleaving tool,*" which has a sharp blade of hard material, such as diamond, sapphire, or tungsten carbide. If proper tension is applied to the fiber as the nick is made, or immediately afterward (this may be done by the cleaving tool in some designs, or manually in other designs), the fracture will propagate in a controlled fashion, creating the desired endface. 2. To break a fiber in such a controlled fashion. *Note:* A good cleave is required for a successful splice of an optical fiber, whether by fusion or mechanical means. Also, some types of fiber-optic connectors do not employ abrasives and polishers. Instead, they use some type of cleaving technique to trim the fiber to its proper length, and produce a smooth, flat perpendicular endface.

client. In networking, a software application that allows the user to access a service from a server computer, *e.g.*, a server computer on the Internet.

client-server. Any hardware/software combination that generally adheres to a client-server architecture, regardless of the type of application.

client-server architecture. Any network-based software system that uses client software to request a specific service, and corresponding server software to provide the service from another computer on the network.

clipper. A circuit or device that limits the instantaneous output signal amplitude to a predetermined maximum value, regardless of the amplitude of the input signal.

clipper chip. An IC designed for secure communications.

clipping. 1. In telephony, the loss of the initial or final parts of a word, words, or syllable, usually caused by the nonideal operation of voice-actuated devices. 2. The limiting of instantaneous signal amplitudes to a predetermined maximum value. 3. In a display device, the removal of those parts of display elements that lie outside of a given boundary.

clock. 1. A reference source of timing information. 2. A device providing signals used in a transmission system to control the timing of certain functions such as the duration of signal elements or the sampling rate. 3. A device that generates periodic, accurately spaced signals used for such purposes as timing, regulation of the operations of a processor, or generation of interrupts.

clock error. The difference between local clock time or value and a designated reference clock time or value. *Note:* Subtracting the clock difference from the local clock brings the local clock into agreement with the reference clock.

clock phase slew. The rate of relative phase change between a given clock signal and a stable reference signal. *Note:* The two signals are generally at or near the same frequency or have an integral multiple frequency relationship.

clock rate. The rate at which a clock issues timing pulses. *Note:* Clock rates are usually expressed in pulses per second, such as 4.96 Mpps (megapulses per second).

clock tolerance. The maximum permissible departure of a clock indication from a designated time reference such as Coordinated Universal Time (UTC).

clock track. A track on which a pattern of signals is recorded to provide a timing reference.

clockwise polarized wave. *Synonym* right-hand (or clockwise) polarized wave.

closed captioning. In broadcast and cable television, the insertion, into the blank lines between frames, of information that may be decoded and displayed on the screen as written words corresponding to those being spoken and transmitted via the conventional audio subcarrier. *Note:* Closed captioning, developed for the hearing-impaired, requires a special decoder, which may be external to, or built into, the television receiver. Closed captioning is mandated by the Americans with Disabilities Act of 1990.

closed circuit. 1. In radio and television transmission, pertaining to an arrangement in which programs are directly transmitted to specific users and not broadcast to the general public. 2. In telecommunications, a circuit dedicated to specific users. 3. A completed electrical circuit.

closed-loop noise bandwidth. The integral, over all frequencies, of the absolute value of the closed-loop transfer function of a phase-locked loop. *Note:* The closed-loop noise bandwidth, when multiplied by the noise spectral density, gives the output noise power in a phase-locked loop.

closed-loop transfer function. A mathematical expression (algorithm) describing the net result of the effects of a closed (feedback) loop on the input signal to the circuits enclosed by the loop. *Note 1:* The closed-loop transfer function is measured at the output. *Note 2:* The output signal waveform can be calculated from the closed-loop transfer function and the input signal waveform.

closed user group. In a network, a group of users permitted to communicate with each other but not with users outside the group. *Note:* A user data terminal equipment (DTE) may belong to more than one closed user group.

closed user group with outgoing access. A closed user group in which at least one member of the group has a facility that permits communication with one or more users external to the closed user group.

closed waveguide. An electromagnetic waveguide (a) that is tubular, usually with a circular or rectangular cross section, (b) that has electrically conducting walls, (c) that may be hollow or filled with a dielectric material, (d) that can support a large number of discrete propagating modes, though only a few may be practical, (e) in which each discrete mode defines the propagation constant for that mode, (f) in which the field at any point is describable in terms of the supported modes, (g) in which there is no radiation field, and (h) in which discontinuities and bends cause mode conversion but not radiation.

closure. *Synonym* splice closure.

cloud attenuation. In the transmission of electromagnetic signals, attenuation caused by absorption and scattering by water or ice particles in clouds. *Note:* The amount of cloud attenuation depends on many factors, including (a) the density, particle size, and turbulence of the clouds and (b) the transmission path length in the clouds.

C-message weighting. A noise spectral weighting used in a noise power measuring set to measure noise power on a line that is terminated by a 500-type set or similar instrument. *Note:* The instrument is calibrated in dBrnC.

CMRR. *Abbreviation for* common-mode rejection ratio.

CNR. *Abbreviation for* carrier-to-noise ratio, combat net radio.

CNS. *Abbreviation for* complementary network services.

C.O.. *Abbreviation for* central office.

COAM. *Acronym for* customer owned and maintained equipment. *Deprecated term. See* customer premises equipment.

coast Earth station. An Earth station in the fixed-satellite service or, in some cases, in the maritime mobile-satellite service, located at a specified fixed point on land to provide a feeder link for the maritime mobile-satellite service.

coasting mode. In timing-dependent systems, a free-running operational timing mode in which continuous or periodic measurement of clock error, *i.e.,* of timing error, is not made. *Note:* Operation in the coasting mode may be enhanced for a period of time by using clock-error data or clock-correction data (obtained during a prior period of operation in the tracking mode) to estimate clock corrections.

coast station. A land station in the maritime mobile service.

coating. *See* primary coating.

coax. *See* coaxial cable.

coaxial cable (coax). A cable consisting of a center conductor surrounded by an insulating material and a concentric outer conductor. *Note:* Coaxial cable is used primarily for wideband, video, or rf applications.

coaxial patch bay. *See* patch bay.

COBOL. *Acronym for* common business oriented language. A programming language designed for business data processing.

co-channel interference. Interference resulting from two or more simultaneous transmissions on the same channel.

code. 1. A set of unambiguous rules specifying the manner in which data may be represented in a discrete form. *Note 1:* Codes may be used for brevity or security. *Note 2:* Use of a code provides a means of converting information into a form suitable for communications, processing, or encryption. 2. [Any] system of communication in which arbitrary groups of letters, numbers, or symbols represent units of plain text of varying length. *Note:* Codes may or may not provide security. Common uses include: (a) converting information into a form suitable for communications or encryption, (b) reducing the length of time required to transmit information, (c) describing the instructions which control the operation of a computer, and (d) converting plain text to meaningless combinations of letters or numbers and vice versa. 3. A cryptosystem in which the cryptographic equivalents, (usually called "code groups") typically consisting of letters or digits (or both) in otherwise meaningless combinations, are substituted for plain text elements which are primarily words, phrases, or sentences. 4. A set of rules that maps the elements of one set, the coded set, onto the elements of another set, the code element set. *Synonym* coding scheme. 5. A set of items, such as abbreviations, that represents corresponding members of another set. 6. To represent data or a computer program in a symbolic form that can be accepted by a processor. 7. To write a routine.

codec. *Acronym for* coder-decoder. 1. An assembly consisting of an encoder and a decoder in one piece of equipment. 2. A circuit that converts analog signals to digital code and vice versa. 3. An electronic device that converts analog signals, such as video and voice signals, into digital form and compresses them to conserve bandwidth on a transmission path. *Note:* Codecs in this sense are used in this sense for video conferencing systems.

code character. A character that (a) is used to represent a discrete value or symbol and (b) is derived in accordance with a code.

code conversion. 1. Conversion of signals, or groups of signals, in one code into corresponding signals, or groups of signals, in another code. 2. A process for converting a code of some predetermined bit structure, such as 5, 7, or 14 bits per character interval, to another code with the same or a different number of bits per character interval. *Note:* In code conversion, alphabetical order is not significant.

coded character set. A character set established in accordance with unambiguous rules that define the character set and the one-to-one relationships between the characters of the set and their coded representations.

coded image. A representation of a display image in a form suitable for storage and processing.

code-division multiple access (CDMA). A coding scheme, used as a modulation technique, in which multiple channels are independently coded for transmission over a single wideband channel. *Note 1:* In some communication systems, CDMA is used as an access method that permits carriers from different stations to use the same transmission equipment by using a wider bandwidth than the individual carriers. On reception, each carrier can be distinguished from the others by means of a specific modulation code, thereby allowing for the reception of signals that were originally overlapping in frequency and time. Thus, several transmissions can occur simultaneously within the same bandwidth, with the mutual interference reduced by the degree of orthogonality of the unique codes used in each transmission. *Note 2:* CDMA permits a more uniform distribution of energy in the emitted bandwidth.

coded set. A set of elements onto which another set of elements has been mapped according to a code. *Note:* Examples of coded sets include the list of names of airports that is mapped onto a set of corresponding three-letter representations of airport names, the list of classes of emission that is mapped onto a set of corresponding standard symbols, and the names of the months of the year mapped onto a set of two-digit decimal numbers.

code element. One of a set of parts of which the characters in a given code may be composed.

code-excited linear prediction (CELP). An analog-to-digital voice coding scheme.

code group. [A] group of letters, numbers, or both in a code system used to represent a plain text word, phrase, or sentence. *Note:* Code groups may include symbols and other elements.

code-independent data communication. *Synonym* code-transparent data communication.

code restriction. A service feature by which certain terminals are prevented from accessing certain features of the network.

code set. The complete set of representations defined by a particular code and language.

code-transparent data communication. A mode of data communication that uses protocols that do not depend for their correct functioning on the data character set or data code used. *Synonym* code-independent data communication.

code word. 1. In a code, a word that consists of a sequence of symbols assembled in accordance with the specific rules of the code and assigned a unique meaning. *Note:* Examples of code words are error-detecting-or-correcting code words and communication code words, such as SOS, MAYDAY, ROGER, TEN-FOUR, and OUT. 2. A cryptonym used to identify sensitive intelligence data. 3. A word that has been assigned a classification and a classified meaning to safeguard intentions and information regarding a classified plan or operation.

coding. 1. In communications systems, the altering of the characteristics of a signal to make the signal more suitable for an intended application, such as optimizing the signal for transmission, improving transmission quality and fidelity, modifying the signal spectrum, increasing the information content, providing error detection and/or correction, and providing data security. *Note:* A single coding scheme usually does not provide more than one or two specific capabilities. Different codes have different advantages and disadvantages. 2. In communications and computer systems, implementing rules that are used to map the elements of one set onto the elements of another set, usually on a one-to-one basis. 3. The digital encoding of an analog signal and, conversely, decoding to an analog signal.

coding scheme. *Synonym* code (def. #4).

codress message. In military communications systems, a message in which the entire address is encrypted with the message text.

COG. *Abbreviation for* centralized ordering group.

coherence area. Pertaining to an electromagnetic wave, the area of a surface perpendicular to the direction of propagation, over which the electromagnetic wave maintains a specified degree of coherence. *Note:* The specified degree of coherence is usually taken to be 0.88 or greater.

coherence degree. *See* degree of coherence.

*****coherence length.** The propagation distance from a coherent source to a point where an electromagnetic wave maintains a specified degree of coherence. *Note 1:* In long-distance transmission systems, the coherence length may be reduced by propagation factors such as dispersion, scattering, and diffraction. *Note 2:* In optical communications, the coherence length, *L*, is given approximately by *L* = ë^2/(*n*Äë), where ë is the central wavelength of the source, *n* is the refractive index of the medium, and Äë is the spectral width of the source. *Note 3: Coherence length* is usually applied to the optical regime.

*****coherence time.** For an electromagnetic wave, the time over which a propagating wave may be considered coherent. *Note 1:* In long-distance transmission systems, the coherence time may be reduced by propagation factors such as dispersion, scattering, and diffraction. *Note 2:* In optical communications, coherence time, ô, is calculated by dividing the coherence length by the phase velocity of light in a medium; approximately given by ô = ë^2/(cÄë) where ë is the central wavelength of the source, Äë is the spectral width of the source, and c is the velocity of light in vacuum. *Note 3:* "*Coherence time*" is usually applied to the optical regime.

coherent. Pertaining to a fixed phase relationship between corresponding points on an electromagnetic wave. *Note:* A truly coherent wave would be perfectly coherent at all points in space. In practice, however, the region of high coherence may extend over only a finite distance.

coherent bundle. *Synonym* aligned bundle.

coherent differential phase-shift keying (CDPSK). Phase-shift keying (a) that is used for digital transmission, (b) in which the phase of the carrier is discretely modulated in relation to the phase of a reference signal and in accordance with data to be transmitted, and (c) in which the modulated carrier is of constant amplitude and frequency. *Note:* A phase comparison is made of successive pulses, and information is recovered by examining the phase transitions between the carrier and successive pulses rather than by the absolute phases of the pulses.

coherent pulse operation. In pulsed carrier transmission, a method of operation in which a fixed phase relationship of the carrier wave is maintained from one pulse to the next.

coherent radiation. *See* coherent.

cold standby. Pertaining to spare electronic equipment that is available for substitute use, but is not powered or warmed up and ready for use.

collective address. *Synonym* group address.

collective routing. Routing in which a switching center automatically delivers messages to a specified list of destinations. *Note 1:* Collective routing avoids the need to list each single address in the message heading. *Note 2:* Major relay stations usually transmit messages bearing collective-routing indicators to tributary, minor, and other major relay stations.

collimation. The process by which a divergent or convergent beam of electromagnetic radiation is converted into a beam with the minimum divergence or convergence possible for that system (ideally, a bundle of parallel rays).

collimator. A device that renders divergent or convergent rays more nearly parallel. *Note:* The degree of collimation (parallelism) should be stated.

collinear antenna array. An array of dipole antennas mounted in such a manner that every element of each antenna is in an extension, with respect to its long axis, of its counterparts in the other antennas in the array. *Note:* A collinear array is usually mounted vertically, in order to increase overall gain and directivity in the horizontal direction. When stacking dipole antennas in such a fashion, doubling their number will, with proper phasing, produce a 3-dB increase in directive gain.

collision. 1. In a data transmission system, the situation that occurs when two or more demands are made simultaneously on equipment that can handle only one at any given instant. 2. In a computer, the situation that occurs when an attempt is made to store simultaneously two different data items at a given address that can hold only one of the items.

color errors. In video systems, distortion of hues in all or a portion of the received image.

color keying. *Synonym* chroma keying.

combat-net radio (CNR). A radio operating in a network that (a) provides a half-duplex circuit and (b) uses either a single radio frequency or a discrete set of radio frequencies when in a frequency hopping mode. *Note:* CNRs are primarily used for push-to-talk-operated radio nets for command and control of combat, combat support, and combat service support operations among ground, sea, and air forces.

combinational logic element. A device having at least one output channel and one or more input channels, all characterized by discrete states, such that at any instant the state of each output channel is completely determined by the states of the input channels at the same instant.

combined communications. The common use of communications facilities by two or more military services, each belonging to a different nation. *Note:* Such use might be specified by a combined communications-electronics agency.

combined distribution frame (CDF). A distribution frame that combines the functions of main and intermediate distribution frames and contains both vertical and horizontal terminating blocks. *Note 1:* The vertical blocks are used to terminate the permanent outside lines entering the station. Horizontal blocks are used to terminate inside plant equipment. This arrangement permits the association of any outside line with any desired terminal equipment. These connections are made either with twisted pair wire, normally referred to as jumper wire, or with optical fiber cables, normally referred to as jumper cables. *Note 2:* In technical control facilities, the vertical side may be used to terminate equipment as well as outside lines. The horizontal side is then used for jackfields and battery terminations.

combined station. In high-level data-link control (HDLC) operation, the station that is usually responsible for performing balanced link-level operations and that generates commands, interprets responses, interprets received commands, and generates responses.

combiner. *See* maximal-ratio combiner.

COMINT. *Acronym for* communications intelligence.

comma-free code. A code constructed so that any partial code word, beginning at the start of a code word but terminating prior to the end of that code word, is not a valid code word. *Note 1:* The comma-free property permits the proper framing of transmitted code words when (a) external synchronization is provided to identify the start of the first code word in a sequence of code words and (b) no uncorrected errors occur in the symbol stream. *Note 2:* Examples of comma-free are the variable-length Huffman codes. *Synonym* prefix-free code.

command. 1. An order for an action to take place. 2. In data transmission, an instruction sent by the primary station instructing a secondary station to perform some specific function. 3. In signaling systems, a control signal. 4. In computer programming, that part of a computer instruction word that specifies the operation to be performed. 5. *Loosely,* a mathematical or logic operator.

command and control (C^2). The exercise of authority and direction by a properly designated commander over assigned and attached forces in the accomplishment of the mission. Command and control functions are performed through an arrangement of personnel, equipment, communications, facilities, and procedures employed by a commander in planning, directing, coordinating, and controlling forces and operations in the accomplishment of the mission.

command and control (C^2) system. The facilities, equipment, communications, procedures, and personnel essential to a commander for planning, directing, and controlling operations of assigned forces pursuant to the missions assigned.

command and control warfare (C^2W). The integrated use of operations security (OPSEC), military deception, psychological operations (PSYOP), electronic warfare (EW), and physical destruction, mutually supported by intelligence, to deny information to, influence, degrade, or destroy adversary command and control capabilities, while protecting friendly command and control capabilities against such actions. *Note:* Command and control warfare applies across the operational continuum and all levels of conflict. C^2W is both offensive and defensive:

(a) **counter-C^2**: To prevent effective C^2 of adversary forces by denying information to, influencing, degrading, or destroying the adversary C^2 systems.

(b) **C^2-protection.** To maintain effective command and control of own forces by turning to friendly advantage or negating adversary efforts to deny information to, influence, degrade, or destroy the friendly C^2 system.

command, control and communications (C^3). The capabilities required by commanders to exercise command and control of their forces.

command, control, communications, computers, and intelligence (C^4I). The facilities, computer equipment, communications equipment, display devices, and intelligence systems necessary to support military operations.

command, control, communications, and computer systems (C^4S). Integrated systems of doctrine, procedures, organizational structures, personnel, equipment, facilities, and communications designed to support a commander's exercise of command and control, through all phases of the operational continuum. *Synonym* C^4 systems.

command frame. In data transmission, a frame, containing a command, transmitted by a primary station.

command menu. A list of all the different commands that may be given to a computer or communications system by an operator. *Note:* Commands on a command menu may be selected by the operator by (a) using an electromechanical pointer, such as a light pen, (b) touching the display screen with a finger, (c) speaking to a voice-recognition system, or (d) positioning a cursor or reverse-video bar by using a keyboard or mouse, and depressing one or more keys on the keyboard or mouse.

command net. A communications network which connects an echelon of command with some or all of its subordinate echelons for the purpose of command and control.

command protocol data unit. A protocol data unit (PDU) transmitted by a logical link control (LLC) sublayer in which the PDU command/response (C/R) bit is equal to "0".

comm center. *Synonym* communications center.

commercial carrier. *Synonym* common carrier.

commercial refile. In military communications systems, the processing of a message from (a) a given military network, such as a tape-relay network, a point-to-point telegraph network, a radio-telegraph network, or the DSN to (b) a commercial communications network. *Note:* Commercial refiling of a message will usually require a reformatting of the message, particularly the heading.

commit transaction. The application, *i.e.,* insertion, of information into any data repository of an integrated database management system in a distributed local communications network.

commonality. 1. A quality that applies to materiel or systems: (a) possessing like and interchangeable characteristics enabling each to be utilized, or operated and maintained by personnel trained on the others without additional specialized training; (b) having interchangeable repair parts and/or components; (c) applying to consumable items interchangeably equivalent without adjustment. 2. Pertaining to equipment or systems that have the quality of one entity possessing like and interchangeable parts with another equipment or system entity. 3. Pertaining to system design in which a given part can be used in more than one place in the system, *i.e.,* subsystems and components have parts in common. *Note:* Examples of commonality include the use of a firing pin that fits in many different weapons and the use of a light source that fits in many different types of fiber optic transmitters.

common battery. A single electrical power source used to energize more than one circuit, component, equipment, or system. *Note 1:* A common battery is usually an electrolytic device and is usually centrally located to the equipment that it serves. *Note 2:* In many telecommunications applications, the common battery is at a nominal –48 vdc. *Note 3:* A central office common battery supplies power to operate all directly connected instruments. *Note 4: Common battery* may include one or more power conversion devices to transform commercial power to direct current, with an electrolytic battery floating across the output.

common-battery signaling. Signaling in which the signaling power of a telephone is supplied by the serving switch. *Note:* In common-battery signaling, "talking power" may be supplied by common or local battery.

common carrier. In a telecommunications context, a telecommunications company that holds itself out to the public for hire to provide communications transmission services. *Note:* In the United States, such companies are usually subject to regulation by Federal and state regulatory commissions. *Synonyms* carrier, commercial carrier, communications common carrier.

common-channel interoffice signaling (CCIS). In multichannel switched networks, a method of transmitting all signaling information for a group of trunks by encoding it and transmitting it over a separate channel using time-division digital techniques.

common-channel signaling. In a multichannel communications system, signaling in which one channel in each link is used for signaling to control, account for, and manage traffic on all channels of the link. *Note:* The channel used for common-channel signaling does not carry user information.

common control. An automatic switching arrangement in which the control equipment necessary for the establishment of connections is shared by being associated with a given call only during the period required to accomplish the control function for the given call. *Note:* In common control, the channels that are used for signaling, whether frequency bands or time slots, are not used for message traffic.

common control switching arrangement (CCSA). An arrangement in which switching for a private network is provided by one or more common control switching systems. *Note:* The switching systems may be shared by several private networks and also may be shared with the public telephone networks.

common control system. An automatic switching system that makes use of common equipment to establish a connection. *Note:* The common equipment then becomes available to establish other connections.

common equipment. Equipment used by more than one system, subsystem, component, or other equipment, such as a channel or switch.

Common Management Information Protocol (CMIP). A protocol used by an application process to exchange information and commands for the purpose of managing remote computer and communications resources.

common management information service (CMIS). A service that specifies the service interface to the Common Management Information Protocol (CMIP). *Note:* To transfer management information between open systems using CMIS/CMIP, peer connections, *i.e.*, associations, must be established. This requires the establishment of an Application Layer association, a Session Layer connection, a Transport Layer connection, and, depending on supporting communications technology, Network Layer and Link Layer connections.

common-mode interference. 1. Interference that appears between signal leads, or the terminals of a measuring circuit, and ground. 2. A form of coherent interference that affects two or more elements of a network in a similar manner (*i.e.*, highly coupled) as distinct from locally generated noise or interference that is statistically independent between pairs of network elements.

common-mode rejection ratio (CMRR). The ratio of the common-mode interference voltage at the input of a circuit, to the corresponding interference voltage at the output.

common-mode voltage. 1. The voltage common to both input terminals of a device. 2. In a differential amplifier, the unwanted part of the voltage between each input connection point and ground that is added to the voltage of each original signal. *Synonym* longitudinal voltage.

common return. A return path that is common to two or more circuits and that serves to return currents to their source or to ground.

common return offset. In a line or circuit, the dc potential difference between ground and the common return.

common user. In communications systems, pertaining to communications facilities and services provided to essentially all users in the area served by the system, rather than to one user or to a relatively small number of users, such as a closed user group with outgoing access.

common user circuit. A circuit designated to furnish a communication service to a number of users.

common user network. A system of circuits or channels allocated to furnish communication paths between switching centers to provide communication service on a common basis to all connected stations or subscribers. It is sometimes described as a general purpose network.

communications. 1. Information transfer, among users or processes, according to agreed conventions. 2. The branch of technology concerned with the representation, transfer, interpretation, and processing of data among persons, places, and machines. *Note:* The meaning assigned to the data must be preserved during these operations.

communications blackout. 1. A cessation of communications or communications capability caused by a lack of power to a communications facility or equipment. 2. A total lack of communications capability caused by propagation anomalies, *e.g.,* those present during strong auroral activity or during the re-entry of a spacecraft into the Earth's atmosphere.

communications center. 1. An agency charged with the responsibility for handling and controlling communications traffic. The center normally includes message center, transmitting, and receiving facilities. 2. A facility that (a) serves as a node for a communications network, (b) is equipped for technical control and maintenance of the circuits originating, transiting, or terminating at the node, (c) may contain message-center facilities, and (d) may serve as a gateway. *Synonym* comm center.

communications channel. *See* channel.

communications common carrier. *Synonym* common carrier.

communications deception. 1. Deliberate transmission, retransmission, or alteration of communications to mislead an adversary's interpretation of the communications. 2. Use of devices, operations, and techniques with the intent of confusing or misleading the user of a communications link or a navigation system.

communications-electronics (C-E). The specialized field concerned with the use of electronic devices and systems for the acquisition or acceptance, processing, storage, display, analysis, protection, disposition, and transfer of information. *Note:* C-E includes the wide range of responsibilities and actions relating to (a) electronic devices and systems used in the transfer of ideas and perceptions, (b) electronic sensors and sensory systems used in the acquisition of information devoid of semantic influence, and (c) electronic devices and systems intended to allow friendly forces to operate in hostile environments and to deny to hostile forces the effective use of electromagnetic resources.

communications intelligence (COMINT). Technical and intelligence information derived from foreign communications by other than the intended recipients.

communications jamming (COMJAM). 1. The portion of electronic jamming that is directed against communications circuits and systems. 2. The prevention of successful radio communications by the use of electromagnetic signals, *i.e.,* the deliberate radiation, reradiation, or reflection of electromagnetic energy with the objective of impairing the effective use of electronic communications systems. *Note:* The aim of communications jamming is to prevent communications by electromagnetic means, or at least to degrade communications sufficiently to cause delays in transmission and reception. Jamming may be used in conjunction with deception to achieve an overall electronic-countermeasure (ECM) plan implementation.

communications net. An organization of stations capable of direct communication on a common channel or frequency. *Synonym* net.

communications net operation. *See* net operation.

communications network. An organization of stations capable of intercommunications but not necessarily on the same channel.

communications processor unit (CPU). A computer embedded in a communications system. *Note 1:* An example of a CPU is the message data processor of a DDN switching center. *Note 2: CPU* is also an abbreviation for *central processing unit* of a computer.

communications protection. The application of communications security (COMSEC) measures to telecommunications systems in order (a) to deny unauthorized persons access to sensitive unclassified information of value, (b) to prevent disruption of telecommunications services, or (c) to ensure the authenticity of information handled by telecommunications systems.

communications protocol. *See* protocol.

communications reliability. The probability that information transmitted from a communications station will arrive at the intended destination in a timely manner without loss of content.

communications satellite. An orbiting vehicle that relays signals between (a) terrestrial communications stations, (b) a terrestrial communications station and another communications satellite, or (c) other communications satellites.

communications saturation. *See* saturation.

communications security (COMSEC). Measures and controls taken to deny unauthorized persons information derived from telecommunications and ensure the authenticity of such telecommunications. *Note:* Communications security includes cryptosecurity, transmission security, emission security, and physical security of COMSEC material.

(a) **cryptosecurity.** [The] component of communications security that results from the provision of technically sound cryptosystems and their proper use.

(b) **emission security.** Protection resulting from all measures taken to deny unauthorized persons information of value which might be derived from intercept and analysis of compromising emanations from crypto-equipment, AIS, and telecommunications systems.

(c) **physical security.** The component of communications security that results from all physical measures necessary to safeguard classified equipment, material, and documents from access thereto or observation thereof by unauthorized persons.

(d) **transmission security.** [The] component of communications security that results from the application of measures designed to protect transmissions from interception and exploitation by means other than cryptanalysis.

communications security equipment. *See* COMSEC equipment.

communications security material. *See* COMSEC material.

communications silence. The avoidance of any type of transmission, emission, or radiation by any means, including radiation from receiving equipment. *Note:* An example of communications silence is the maintaining of a listening watch only if the receivers do not radiate beyond a specified level.

communications sink. *See* sink.

communications source. *See* source.

communications subsystem. A functional unit or operational assembly that is smaller than the larger assembly under consideration. *Note:* Examples of communications subsystems in the Defense Communications System (DCS) are (a) a satellite link with one Earth terminal in CONUS and one in Europe, (b) the interconnect facilities at each Earth terminal of the satellite link, and (c) an optical fiber cable with its driver and receiver in either of the interconnect facilities.

communications survivability. The ability of communications systems to continue to operate effectively under adverse conditions, though portions of the system may be damaged or destroyed. *Note:* Various methods may be used to maintain communications services, such as using alternate routing, different transmission media or methods, redundant equipment, and sites and equipment that are radiation hardened.

communications system. A collection of individual communications networks, transmission systems, relay stations, tributary stations, and data terminal equipment (DTE) usually capable of interconnection and interoperation to form an integrated whole. *Note:* The components of a communications system serve a common purpose, are technically compatible, use common procedures, respond to controls, and operate in unison.

communications system engineering. The translation of user requirements for the exchange of information into cost-effective and low-risk technical solutions in terms of equipment and subsystems. *Note:* Communications system engineering encompasses the integration of these parts into a complete entity resulting in a minimum investment for the entire system lifecycle required to satisfy the requirements of a majority of users of the communication system.

communications system survivability. *See* survivability.

communications theory. Theory that is devoted to the probabilistic characteristics of the transmission of data in the presence of noise, and that is used to advance the design, development, and operation of communications systems.

communications watch. The monitoring of one or more communications lines, frequencies, or channels to obtain information by listening to or receiving all transmissions on them and transmitting and receiving messages as required.

communications zone. [The] rear part of theater of operations (behind but contiguous to the combat zone), which contains the lines of communications, establishments for supply and evacuation, and other agencies required for the immediate support and maintenance of the field forces.

community antenna television (CATV). *See* cable TV.

community of interest. A grouping of users who generate a majority of their traffic in calls to other members of the group. *Note:* The community of interest may be related to a geographic area or to an administrative organization. *Synonym* special interest group.

community reception (in the broadcasting-satellite service). The reception of emissions from a space station in the broadcasting-satellite service by receiving equipment, which in some cases may be complex and have antennae larger than those used for individual reception, and intended for use:

- by a group of the general public at one location;

or

- through a distribution system covering a limited area.

compact. *See* data compaction.

compact disk read-only memory. *See* CD ROM.

compaction. *See* data compaction.

companding. An operation in which the dynamic range of signals is compressed before transmission and is expanded to the original value at the receiver. *Note:* The use of companding allows signals with a large dynamic range to be transmitted over facilities that have a smaller dynamic range capability. Companding reduces the noise and crosstalk levels at the receiver.

compandor. A device that incorporates a compressor and an expander, each of which may be used independently.

comparably efficient interconnection (CEI). An equal-access concept developed by the FCC stating that, "... if a carrier offers an enhanced service, it should be required to offer network interconnection (or collocation) opportunities to others that are comparably efficient to the interconnection that its enhanced service enjoys. Accordingly, a carrier would be required to implement CEI only as it introduces new enhanced services." [FCC *Report and Order* June 16, 1986]

comparator. 1. In analog computing, a functional unit that compares two analog variables and indicates the result of that comparison. 2. A device that compares two items of data and indicates the result of that comparison. 3. A device for determining the dissimilarity of two items such as two pulse patterns or words.

compartmentation. The segregation of components, programs, and information to provide isolation. *Note:* Compartmentation provides some protection against overall compromise, contamination, or unauthorized access.

compatibility. 1. Capability of two or more items or components of equipment or material to exist or function in the same system or environment without mutual interference. 2. In computing, the ability to execute a given program on different types of computers without modification of the program or the computers. 3. The capability that allows the substitution of one subsystem (storage facility), or of one functional unit (*e.g.*, hardware, software), for the originally designated system or functional unit in a relatively transparent manner, without loss of information and without the introduction of errors.

compatible sideband transmission. Independent sideband transmission in which the carrier is deliberately reinserted at a lower level after its normal suppression to permit reception by conventional AM receivers. *Note:* Compatible sideband transmission is usually single-sideband (SSB) amplitude-modulation-equivalent (AME) transmission consisting of the emission of the carrier plus the upper sideband. *Synonym* amplitude modulation equivalent.

compelled signaling. Signaling in which the transmission of each signal in the forward direction from an originating terminal is inhibited until an acknowledgement of the satisfactory receipt of the previous signal is received by the originating terminal.

competitive access provider (CAP). A company that provides exchange access services in competition with an established U.S. telephone local exchange carrier.

competitive clip. In time-assignment speech interpolation (TASI) or digital speech interpolation (DSI), truncation of the initial part of a speech spurt, caused when all channels in a given direction of transmission are busy and the transmission of the spurt must wait for an available channel.

compile. 1. To translate a computer program expressed in a high-level language into a program expressed in a lower level language, such as an intermediate language, assembly language, or a machine language. 2. To prepare a machine language program from a computer program written in another programming language by making use of the overall logic structure of the program or by generating more than one computer instruction for each symbolic statement as well as performing the function of an assembler.

compiler. A computer program for compiling. *Synonym* compiling program.

compiling program. *Synonym* compiler.

complementary network service (CNS). A means for an enhanced service provider customer to connect to a network and to the enhanced service provider. *Note:* Complementary network services usually consist of the customer local service, such as a business or residence, and several associated service features, such as a call-forwarding service.

component. 1. An assembly, or part thereof, that is essential to the operation of some larger assembly and is an immediate subdivision of the assembly to which it belongs. *Note:* For example, a radio receiver may be a component of a complete radio set consisting of a combined transmitter-receiver, *i.e.,* a transceiver. The same radio receiver could also be a subsystem of the combined transmitter-receiver, in which case the IF amplifier section would be a component of the receiver but not of the radio set. Similarly, within the IF amplifier section, items, such as resistors, capacitors, vacuum tubes, and transistors, are components of that section. 2. In logistics, a part, or combination of parts having a specified function, that can only be installed or replaced as an entity. . 3. In material, an assembly or any combination of parts, subassemblies, and assemblies mounted together in manufacture, assembly, maintenance, or rebuild.

composite cable. A communications cable having both optical and metallic signal-carrying components. *Note 1:* A cable having optical fiber(s) and a metallic component, *e.g.,* a metallic twisted pair, used solely for conduction of electric power to repeaters, does qualify as a composite cable. *Note 2:* A cable having optical fiber(s) , plus a metallic strength member or armor, does not qualify as a composite cable. *Contrast with* hybrid cable.

composited circuit. A circuit that can be used simultaneously either for telephony and dc telegraphy or for telephony and signaling. *Note:* Separation of the two may be accomplished by frequency discrimination. *Synonym* voice-plus circuit.

composite signaling (CX). Signaling in which an arrangement is made to provide direct current signaling and dial pulsing beyond the range of conventional loop signaling. *Note:* Composite signaling, like DX signaling, permits duplex operation, *i.e.*, permits simultaneous two-way signaling. *Synonym* CX signaling.

composite two-tone test signal. A test signal composed of two different frequencies and used for intermodulation distortion measurements.

composite video. In television, a video signal in which synchronizing information (pulses) and picture information, including chroma, *i.e.*, color, information are combined.

compound signal. In ac signaling, a signal consisting of the simultaneous transmission of more than one frequency. *Note:* An example of compound signaling is dual-tone multifrequency (DTMF) signaling.

compress. *See* data compaction, data compression, signal compression.

compressed video. Video that has been encoded so as to reduce the number of bits required for storage or transmission.

compression. *See* data compression, signal compression.

compression ratio. 1. In signal compression, the ratio of the dynamic range of compressor input signals to the dynamic range of the compressor output signals. *Note:* The compression ratio is usually expressed in dB. For example, a 40-dB input range compressed to a 30-dB output range would be equivalent to a 10-dB compression. 2. In digital facsimile, the ratio of the total pels scanned for the object to the total encoded bits sent for picture information. 3. The ratio of the gain of a device at a low power level to the gain at some higher level. *Note:* The compression ratio is usually expressed in dB.

compressor. A nonlinear analog device that has a lower gain at higher input levels than at lower input levels. *Note:* A compressor is used to allow signals with a large dynamic range to be sent through devices or circuits with a smaller dynamic range .

compromise. 1. The known or suspected exposure of clandestine personnel, installations, or other assets or of classified information or material, to an unauthorized person. 2. The disclosure of cryptographic information to unauthorized persons. 3. The recovery of plain text of encrypted messages by unauthorized persons through cryptanalysis methods. 4. The disclosure of information or data to unauthorized persons, or a violation of the security policy of a system in which unauthorized intentional or unintentional disclosure, modification, destruction, or loss of an object may have occurred.

compromising emanations. Unintentional signals that, if intercepted and analyzed would disclose the information transmitted, received, handled, or otherwise processed by telecommunications or automated systems equipment.

computer. 1. A device that accepts data, processes the data in accordance with a stored program, generates results, and usually consists of input, output, storage, arithmetic, logic, and control units. 2. A functional unit that can perform substantial computation, including numerous arithmetic operations or logic operations, without human intervention during a run.

Note 1: This definition, approved by the Customs Council, distinguishes a computer from similar devices, such as hand-held calculators and certain types of control devices. *Note 2:* Computers have been loosely classified into microcomputers, minicomputers, and main-frame computers, based on their size. These distinctions are rapidly disappearing as the capabilities of even the smaller units have increased. Microcomputers now are usually more powerful and versatile than the minicomputers and the main-frame computers were a few years ago.

computer-aided software engineering. *See* CASE.

computer-aided systems engineering. *See* CASE.

computer architecture. Of a computer, the physical configuration, logical structure, formats, protocols, and operational sequences for processing data, controlling the configuration, and controlling the operations. *Note:* Computer architecture may also include word lengths, instruction codes, and the interrelationships among the main parts of a computer or group of computers.

computer conferencing. 1. Teleconferencing supported by one or more computers. 2. An arrangement in which access, by multiple users, to a common database is mediated by a controlling computer. 3. The interconnection of two or more computers working in a distributed manner on a common application process.

computer-dependent language. *Synonym* assembly language.

computer graphics. 1. Graphics implemented through the use of computers. 2. Methods and techniques for converting data to or from graphic displays via computers. 3. The branch of science and technology concerned with methods and techniques for converting data to or from visual presentation using computers.

computer language. *Synonym* programming language.

computer network. 1. A network of data processing nodes that are interconnected for the purpose of data communication. 2. A communications network in which the end instruments are computers.

computer network engineering. *See* network engineering (def. # 2).

computer network operating system (NOS). A specialized operating system designed for computer networking on minicopmputers and microcomputers in a local networking area / campus area network. *Note:* A NOS is usually designed to run on existing software designed for that computer and may require interface hardware for the workstation and server.

computer-oriented language. A programming language in which words and syntax are designed for use on a specific computer or class of computers. *Synonyms* low-level language, machine-oriented language.

computer peripheral. *See* peripheral equipment.

computer program. *See* program.

computer program origin. The address assigned to the initial storage location of a computer program in main storage.

computer routine. *See* routine.

computer science. The discipline that is concerned with methods and techniques relating to data processing performed by automatic means.

computer security (COMPUSEC). 1. Measures and controls that ensure confidentiality, integrity, and availability of the information processed and stored by a computer. *Synonym* automated information systems security. 2. The application of hardware, firmware and software security features to a computer system in order to protect against, or prevent, the unauthorized disclosure, manipulation, deletion of information or denial of service.

computer system. A functional unit, consisting of one or more computers and associated software, that (a) uses common storage for all or part of a program and also for all or part of the data necessary for the execution of the program, (b) executes user-written or user-designated programs, and (c) performs user-designated data manipulation, including arithmetic and logic operations. *Note:* A computer system may be a stand-alone system or may consist of several interconnected systems. *Synonyms* ADP system, computing system.

computer system fault tolerance. The ability of a computer system to continue to operate correctly even though one or more of its components are malfunctioning. *Note:* System performance, such as speed and throughput, may be diminished until the faults are corrected. *Synonym* computer system resilience.

computer system resilience. *Synonym* computer system fault tolerance.

computer systems engineering. *See* systems design.

computer word. In computing, a group of bits or characters that occupies one or more storage locations and is treated by computers as a unit. *Synonym* machine word.

computing system. *Synonym* computer system.

COMSEC. *Acronym for* communications security.

COMSEC equipment. Equipment designed to provide security to telecommunications by converting information to a form unintelligible to an unauthorized interceptor and, subsequently, by reconverting such information to its original form for authorized recipients; also, equipment designed specifically to aid in, or as an essential element of, the conversion process. *Note:* COMSEC equipment includes crypto-equipment, crypto-ancillary equipment, cryptoproduction equipment, and authentication equipment.

COMSEC material. [An] item designed to secure or authenticate telecommunications. *Note:* COMSEC material includes, but is not limited to, key, equipment, devices, documents, firmware or software that embodies or describes cryptographic logic and other items that perform COMSEC functions.

concentrator. 1. In data transmission, a functional unit that permits a common path to handle more data sources than there are channels currently available within the path. *Note:* A concentrator usually provides communication capability between many low-speed, usually asynchronous channels and one or more high-speed, usually synchronous channels. Usually different speeds, codes, and protocols can be accommodated on the low-speed side. The low-speed channels usually operate in contention and require buffering. 2. A device that connects a number of circuits, which are not all used at once, to a smaller group of circuits for economy.

concentricity error. In an optical fiber, the distance between the center of the two concentric circles that specify the cladding diameter and the center of the two concentric circles that specify the core diameter. *Note:* The concentricity error is used in conjunction with tolerance fields to specify or characterize optical fiber core and cladding geometry. *Synonyms* core eccentricity, core-to-cladding concentricity, core-to-cladding eccentricity, core-to-cladding offset.

concurrent operation. 1. *Synonym* multitasking. 2. In data link operations, the operation in which two or more data links are used during the same, usually short, time interval, while adhering to the protocols of each link without providing data forwarding among the links.

conditioned baseband representation. *Synonym* non-return-to-zero mark.

conditioned circuit. A communications circuit optimized to obtain desired characteristics for voice or data transmission.

conditioned diphase modulation. A form of diphase modulation, combined with signal conditioning, that (a) eliminates the dc component of the signal, (b) enhances timing recovery, and (c) facilitates transmission over voice frequency (VF) circuits or coaxial cables.

conditioned loop. A loop that has conditioning equipment to obtain the desired line characteristics for voice or data transmission. *Note:* The conditioning equipment is used to improve the amplitude-vs.-frequency characteristics of the circuit and to match impedance.

conditioning equipment. 1. At junctions of circuits, equipment used to obtain desired circuit characteristics, such as matched transmission levels, matched impedances, and equalization between facilities. 2. Corrective networks used to improve data transmission, such as equalization of the insertion-loss-vs.-frequency characteristic and the envelope delay distortion over a desired frequency range.

conducted interference. 1. Interference resulting from noise or unwanted signals entering a device by conductive coupling, *i.e.,* by direct coupling. 2. An undesired voltage or current generated within, or conducted into, a receiver, transmitter, or associated equipment, and appearing at the antenna terminals.

conduction band. 1. In a semiconductor, the range of electron energy, higher than that of the valence band, sufficient to make the electrons free to move from atom to atom under the influence of an applied electric field and thus constitute an electric current. 2. In the atomic structure of a material, a partially filled or empty energy level in which electrons are free to move, thus allowing the material to conduct an electrical current upon application of an electric field by means of an applied voltage.

conductive coupling. Energy transfer achieved by means of physical contact, *i.e.,* coupling other than inductive or capacitive coupling. *Note 1:* Conductive coupling may be achieved by wire, resistor, or common terminal, such as a binding post or metallic bonding. *Note 2:* Conductive coupling passes the full spectrum of frequencies, including dc. *Synonym* direct coupling.

cone of silence. *Synonym* antenna blind cone.

CONEX. *Acronym for* connectivity exchange.

conference call. A service feature that allows a call to be established among three or more stations in such a manner that each of the stations is able to communicate with all the other stations. *Synonym* multiple call.

conference operation. In a communications network, operation that allows a call to be established among three or more stations in such a manner that each of the stations is able to communicate directly with all the other stations. *Note:* In radio systems, the stations may receive simultaneously, but must transmit one at a time. The common operational modes are "push-to-talk" for telephone operation and "push-to-type" for telegraph and data transmission.

configuration. In a communications or computer system, an arrangement of functional units according to their nature, number, and chief characteristics. *Note 1:* Configuration pertains to hardware, software, firmware, and documentation. *Note 2:* The configuration will affect system performance.

configuration control. 1. After establishing a configuration, such as that of a telecommunications or computer system, the evaluating and approving changes to the configuration and to the interrelationships among system components. 2. In distributed-queue dual-bus (DQDB) networks, the function that ensures the resources of all nodes of a DQDB network are configured into a correct dual-bus topology. *Note:* The functions that are managed include the head of bus, external timing source, and default slot generator functions.

configuration management. 1. [The] management of security features and assurances through control of changes made to hardware, software, firmware, documentation, test, test fixtures and test documentation of an automated information system, throughout the development and operational life of a system. 2. The control of changes–including the recording thereof–that are made to the hardware, software, firmware, and documentation throughout the system lifecycle.

confirmation signaling. Signaling that ensures error-free transmission of dialed information by returning a unique digit-dependent signal from the far end as each digit is sent over a trunk.

confirmation to receive. In facsimile, a signal from a CCITT Group 1, 2, or 3 facsimile receiver, indicating it is ready to receive picture signals.

conformance test. A test performed by an independent body to determine if a particular piece of equipment satisfies the criteria in a specified controlling document, such as a Federal standard, an American National Standard, a Military Standard, or a Military Specification. *Contrast with* acceptance test.

congestion. 1. In a communications switch, a state or condition that occurs when more subscribers attempt simultaneously to access the switch than it is able to handle, even if unsaturated. 2. In a saturated communications system, the condition that occurs when an additional demand for service occurs.

connecting arrangement. In the public switched telephone networks, the equipment provided by a common carrier to accomplish electrical interconnection between customer-provided equipment and the facilities of the common carrier.

connection. 1. A provision for a signal to propagate from one point to another, such as from one circuit, line, subassembly, or component to another. 2. An association established between functional units for conveying information.

connection-in-progress signal. A call control signal at the data circuit-terminating-equipment/data-terminal-equipment (DCE/DTE) interface that indicates to the DTE that the establishment of the data connection is in progress and that the ready-for-data signal will follow.

connectionless data transfer. *See* connectionless mode transmission.

connectionless mode transmission. In a packet-switched network, transmission in which each packet is encoded with a header containing a destination address sufficient to permit the independent delivery of the packet without the aid of additional instructions. *Note 1:* A packet transmitted in a connectionless mode is frequently called a datagram. *Note 2:* In connectionless mode transmission of a packet, the service provider usually cannot guarantee there will be no loss, error insertion, misdelivery, duplication, or out-of-sequence delivery of the packet. However, the risk of these hazards' occurring may be reduced by providing a reliable transmission service at a higher protocol layer, such as the Transport Layer of the Open Systems Interconnection—Reference Model.

connectionless transmission. *See* connectionless-mode transmission.

connection-mode transmission. *See* connection-oriented mode transmission.

connection-oriented data transfer protocol. A data-transfer protocol in which a logical connection is established between end user terminals.

connection-oriented mode transmission. In a packet-switched network, a mode of transmission in which there is a complete information transfer transaction for each packet or group of packets, *i.e.,* the information transfer phase is preceded by an access phase and followed by a disengagement phase. *Note 1:* During the information transfer phase of connection-oriented mode transmission, more than one packet may be transmitted. The header of each information packet contains a sequence number and an identifier field that associates the packet with the connection that was established during the access phase before the information transfer phase begins. *Note 2:* Connection-oriented mode transmission usually enables detection of lost, erroneous, duplicated, or out-of-sequence packets because a connection is established from end to end before transmission begins. *Note 3:* The CCITT

X.25 protocols are widely used to implement connection-oriented mode transmission on packet-switched public data networks. The protocols are implemented at Layers 1, 2, and 3 of the Open Systems Interconnection—Reference Model.

connections per circuit hour (CCH). 1. A unit of traffic measurement expressed as the number of connections established at a switching point per hour. 2. A unit of traffic measurement used to express the rate at which circuits are established at a switch. *Note:* The magnitude of the CCH is an instantaneous value subject to change as a function of time, *i.e.,* from moment to moment.

connectivity exchange (CONEX). In an adaptive or manually operated high-frequency (HF) radio network, the automatic or manual exchange of information concerning routes to stations that are not directly reachable by the exchange originator. *Note:* The purpose of the exchange is to identify indirect paths and/or possible relay stations to those stations that are not directly reachable.

connector. A device for mating and demating electrical power connections or communications media. *Note:* A connector is distinguished from a splice, which is a permanent joint.

conservation of radiance. A basic principle of optics, that no passive optical system can increase the quantity L/n^2, where L is the radiance of a beam and n is the local refractive index. *Note: "Conservation of radiance"* was formerly called *"conservation of brightness"* or the *"brightness theorem."*

consolidated local telecommunications service. Local communications service provided by GSA to all Federal agencies located in a building, complex, or geographical area.

consultation hold. A service feature that allows a speaker on an extension instrument to place one call on hold and to speak with another caller on a separate line.

content-addressable storage. *Synonym* associative storage (def. #1).

contention. 1. A condition that arises when two or more data stations attempt to transmit at the same time over a shared channel, or when two data stations attempt to transmit at the same time in two-way alternate communication. *Note:* A contention can occur in data communications when no station is designated a master station. In contention, each station must monitor the signals and wait for a quiescent condition before initiating a bid for master status. 2. Competition by users of a system for use of the same facility at the same time.

continuity check. A check made of a circuit to verify that a communication or power path exists.

continuously variable slope delta (CVSD) modulation. A type of delta modulation in which the size of the steps of the approximated signal is progressively increased or decreased as required to make the approximated signal closely match the input analog wave.

continuous operation. 1. Operation in which certain components, such as nodes, facilities, circuits, or equipment, are in an operational state at all times. *Note:* Continuous operation usually requires that there be fully redundant configuration, or at least a sufficient *X* out of *Y* degree of redundancy for compatible equipment, where *X* is the number of spare components and *Y* is the number of operational components. 2. In data transmission, operation in which the master station need not stop for a reply from a slave station after transmitting each message or transmission block.

continuous presence. In teleconferencing, the concurrent presence of two or more video images, such as two images that may appear on a single monitor on a split screen or on two separate monitors.

continuous tone copy. In facsimile, an object, *i.e.,* an original, or a recorded copy, that contains shades of gray, *i.e.,* contains densities between black and white, such as in a photographic print.

continuous wave (cw). A wave of constant amplitude and constant frequency.

contouring. In digital facsimile, density step lines in the recorded copy resulting from analog-to-digital conversion when the object, *i.e.,* the original, has observable shades of gray between the smallest density steps of the digital system.

contrast. 1. In display systems, the relation between (a) the intensity of color, brightness, or shading of an area occupied by a display element, display group, or display image on the display surface of a display device and (b) the intensity of an area not occupied by a display elements, a display group, or a display image. *Deprecated synonym* brightness ratio. 2. In optical character recognition, the difference between color or shading of the printed material on a document and the background on which it is printed.

contribution. In B-ISDN applications, the use of broadband transmission of audio or video information to the user for post-production processing and distribution.

control ball. *Synonym* trackball.

control character. A character that initiates, modifies, or stops a function, event, operation, or control operation. *Note:* Control characters may be recorded for use in subsequent actions. They are not graphic characters but may have a graphic representation in some circumstances.

control communications. The branch of technology devoted to the design, development, and application of communications facilities used specifically for control purposes, such as for controlling (a) industrial processes, (b) movement of resources, (c) electric power generation, distribution, and utilization, (d) communications networks, and (e) transportation systems.

control field. In a protocol data unit (PDU), the field that (a) contains data interpreted by the receiving destination logical-link controller (LLC) and (b) may be the field immediately following the destination service access point (DSAP) and source service access point (SSAP) address fields of the PDU.

control function. *Synonym* control operation.

controlled access. Access in which the resources of an area or system is limited to authorized personnel, users, programs, processes, or other systems, and denied to all others.

controlled area. 1. An area (a) in which uncontrolled movement will not result in compromise of classified information, (b) that is designed to provide administrative control and safety, or (c) that serves as a buffer for controlling access to limited-access areas. 2. An area to which security controls have been applied to protect an information-processing system's equipment and wirelines, equivalent to that required for the information transmitted through the system.

controlled not-ready signal. A signal, sent in the backward direction, to indicate that a call cannot be completed because the called line is not in a ready condition, but is under control, as opposed to not being in a ready condition and not under control.

controlled security mode. *See* controlled security operation.

controlled security operation. In a communications system, operation in which (a) internal security controls prevent inadvertent disclosures, (b) personnel, physical, and administrative controls are used to prevent unauthorized access, (c) both cleared and uncleared users are serviced, and (d) if required, both secured and unsecured remote terminal areas are serviced.

controlled space. Three-dimensional space surrounding telecommunications and automated information systems equipment, within which unauthorized persons are denied unrestricted access and are either escorted by authorized persons or are under continuous physical or electronic surveillance. *Synonym* restricted area.

controller. In an automated radio, the device that commands the radio transmitter and receiver, and that performs processes, such as automatic link establishment, channel scanning and selection, link quality analysis, polling, sounding, message store and forward, address protection, and anti-spoofing.

control of electromagnetic radiation. 1. Measures taken to minimize unintended electromagnetic radiation emanating from a system or component and to minimize electromagnetic interference. *Note:* Control of electromagnetic radiation is exercised for purposes of security and the reduction of interference, especially on ships and aircraft. 2. A national operational plan to minimize the use of electromagnetic radiation in the United States and its possessions and the Panama Canal Zone in the event of attack or imminent threat thereof, as an aid to the navigation of hostile aircraft, guided missiles, or other devices.

control operation. An operation that affects the recording, processing, transmission, or interpretation of data. *Note:* Examples of control operations include starting and stopping a process; executing a carriage return, a font change, or a rewind; and transmitting an end-of-transmission (EOT) control character. *Synonym* control function.

control station. In a communications network, the station that selects the master station and supervises operational procedures, such as polling, selecting, and recovery. *Note:* The control station has the overall responsibility for the orderly operation of the entire network.

conversational mode. A mode of communication analogous to a conversation between two persons.

conversational service. In telecommunications, a service that provides two-way, interactive, real-time, end-to-end information transfer.

convolutional code. A type of error-correction code in which (a) each m-bit information symbol (each m-bit string) to be encoded is transformed into an n-bit symbol, where $n>m$ and (b) the transformation is a function of the last k information symbols, where k is the constraint length of the code. *Note:* Convolutional codes are often used to improve the performance of radio and satellite links.

convoy internal communications. In a land or maritime convoy, communications (a) that is among the elements of the convoy, (b) includes radio, visual, and sound transmissions, (c) in which radio intervehicle or intership communications are usually by radiotelephone using the receipt method of operation, (d) in which the convoy commander vehicle or ship usually has the net-control station aboard and (e) that usually use a single assigned frequency.

cooperation factor. In facsimile systems, the product of the total scanning length and the scanning density, given by $CF = L\sigma$, where L is the scanning line length and σ is the scanning line density, both in compatible units. *Note:* For example, a 20-cm line and a line density of 6 scanning pitches per centimeter would yield a cooperation factor of 120.

coordinated clock. One of a set of clocks distributed over a spatial region, producing time scales that are synchronized to the time scale of a reference clock at a specified location.

coordinated time scale. A time scale synchronized within given tolerances to a reference time scale.

Coordinated Universal Time (UTC). Time scale, based on the second (SI), as defined and recommended by the CCIR, and maintained by the Bureau International des Poids et Mesures (BIPM). For most practical purposes associated with the *Radio Regulations,* UTC is equivalent to mean solar time at the prime meridian (0° longitude), formerly expressed in GMT. *Note 1:* The maintenance by BIPM includes cooperation among various national laboratories around the world. *Note 2:* The full definition of UTC is contained in CCIR Recommendation 460-4. *Note 3*: The second was formerly defined in terms of astronomical phenomena. When this practice was abandoned in order to take advantage of atomic resonance phenomena ("atomic time") to define the second more precisely, it became necessary to make occasional adjustments in the "atomic" time scale to coordinate it with the workaday mean solar time scale, UT-1, which is based on the somewhat irregular rotation of the Earth. Rotational irregularities usually result in a net decrease in the Earth's average rotational velocity, and ensuing lags of UT-1 with respect to UTC. *Note 4:* Adjustments to the atomic, *i.e.*, UTC, time scale consist of an occasional addition or deletion of one full second, which is called a *leap second.* Twice yearly, during the last minute of the day of June 30 and December 31, Universal Time, adjustments may be made to ensure that the accumulated difference between UTC and UT-1 will not exceed 0.9 s before the next scheduled adjustment. Historically, adjustments, when necessary, have usually consisted of adding an extra second to the UTC time scale in order to allow the rotation of the Earth to "catch up." Therefore, the last minute of the UTC time scale, on the day when an adjustment is made, will have 59 or 61 seconds. *Synonyms* World Time, Z Time, Zulu Time.

coordination area. The area associated with an Earth station outside of which a terrestrial station sharing the same frequency band neither causes nor is subject to interfering emissions greater than a permissible level.

coordination contour. 1. The line enclosing the coordination area. 2. The perimeter of the coordination area.

coordination distance. Distance on a given azimuth from an Earth station beyond which a terrestrial station sharing the same frequency band neither causes nor is subject to interfering emissions greater than a permissible level.

copy. 1. To receive a message. 2. A recorded message or a duplicate of it. 3. To read data from a source, leaving the source data unchanged at the source, and to write the same data elsewhere, though they may be in a physical form that differs from that of the source. 4. To understand a transmitted message.

copy-protected. Of a data medium such as a diskette, pertaining to the use of one or more schemes designed to thwart copying in violation of copyright law or security considerations.

copy watch. A radio- or video-communications watch in which an operator is required to maintain a continuous receiver watch and to keep a complete log.

cord circuit. A switchboard circuit in which a plug-terminated cord is used to establish connections manually between user lines or between trunks and user lines. *Note:* A number of cord circuits are furnished as part of the switchboard position equipment. The cords may be referred to as front cord and rear cord or trunk cord and station cord. In modern cordless switchboards, the cord-circuit function is switch operated and may be programmable.

cord lamp. The lamp associated with a cord circuit that indicates supervisory conditions for the respective part of the connection.

cordless switchboard. A telephone switchboard in which manually operated keys are used to make connections.

core. 1. The central region about the longitudinal axis of an optical fiber, which region supports guiding of the optical signal. *Note 1:* For the fiber to guide the optical signal, the refractive index of the core must be slightly higher than that of the cladding. *Note 2:* In different types of fibers, the core and core-cladding boundary function slightly differently in guiding the signal. Especially in single-mode fibers, a significant fraction of the energy in the bound mode travels in the cladding. 2. A piece of ferromagnetic material, usually toroidal in shape, used as a component in a computer memory device. *Note:* The type of memory referred to has very limited application in today's computer environment. It has been largely replaced by semiconductor and other technologies. 3. The material at the center of an electromechanical relay or solenoid, about which the coil is wound.

core area. The part of the cross-sectional area of an optical fiber within which the refractive index is everywhere greater than that of the innermost homogeneous cladding, by a specified fraction of the difference between the maximum refractive index of the core and the refractive index of the innermost cladding. *Note 1:* Artifacts of the manufacturing process, such as refractive index dip, are ignored in computing the points (refractive indices) of demarcation. *Note 2:* The core area is the cross-sectional area within which the refractive index is given by

$$n_3 > [n_2 + m(n_1 - n_2)],$$

where n_1 is the maximum refractive index of the core, n_2 is the refractive index of the homogeneous cladding adjacent to the core, n_3 is the defining refractive index, and m is a fraction, usually not greater than 0.05.

core-cladding offset. *See* concentricity error.

core diameter. In the cross section of a realizable optical fiber, ideally circular, but assumed to a first approximation to be elliptical, the average of the diameters of the smallest circle that can be circumscribed about the core-cladding boundary, and the largest circle that can be inscribed within the core-cladding boundary.

core dump. A printout, usually in hexadecimal characters, of the contents of core memory. *Note:* A core dump is useful for analyzing an abnormally terminated computer run or finding bugs in a computer program.

core eccentricity. *Synonym* concentricity error. *See* ovality.

core noncircularity. *See* ovality.

core storage. *See* magnetic core storage.

core-to-cladding concentricity. *Synonym* concentricity error.

core-to-cladding eccentricity. *Synonym* concentricity error.

core-to-cladding offset. *Synonym* concentricity error.

corner reflector. 1. A reflector consisting of three mutually perpendicular intersecting conducting flat surfaces, which returns a reflected electromagnetic wave to its point of origin. *Note:* Such reflectors are often used as radar targets. 2. A directional antenna using two mutually intersecting conducting flat surfaces. 3. A device, normally consisting of three metallic surfaces or screens perpendicular to one another, designed to act as a radar target or marker. 4. In radar interpretation, an object that, by means of multiple reflections from smooth surfaces, produces a radar return of greater magnitude than might be expected from the physical size of the object. 5. A passive optical mirror, that consists of three mutually perpendicular flat, intersecting reflecting surfaces, which returns an incident light beam in the opposite direction.

corner reflector antenna. *See* corner reflector (def #2).

corrective maintenance. 1. Maintenance actions carried out to restore a defective item to a specified condition. 2. Tests, measurements, and adjustments made to remove or correct a fault.

cosine emission law. *Synonym* Lambert's cosine law.

cosite. Collocation of electronic equipment on the same vehicle, station, or base. *Note:* Equipment so located is often subject to interference because of its proximity to other equipment.

cosmic noise. Random noise that originates outside the Earth's atmosphere. *Note:* Cosmic noise characteristics are similar to those of thermal noise. Cosmic noise is experienced at frequencies above about 15 MHz when highly directional antennas are pointed toward the Sun or to certain other regions of the sky such as the center of the Milky Way Galaxy. *Synonym* galactic radio noise.

Costas loop. A phase-locked loop used for carrier phase recovery from suppressed-carrier modulation signals, such as from double-sideband suppressed carrier signals. *Note:* In the usual implementation of a Costas loop, a local voltage-controlled oscillator provides quadrature outputs, one to each of two phase detectors, *i.e.,* product detectors. The same phase of the input signal is also applied to both phase detectors and the output of each phase detector is passed through a low-pass filter. The outputs of these low-pass filters are inputs to another phase detector, the output of which passes through a loop filter before being used to control the voltage-controlled oscillator.

Coulomb's law. The universal law of attraction and repulsion of electric charges.

counterpoise. A conductor or system of conductors used as a substitute for earth or ground in an antenna system.

country code. 1. In international direct telephone dialing, a code that consists of 1-, 2-, or 3-digit numbers in which the first digit designates the region and succeeding digits, if any, designate the country. 2. In international record carrier transmissions, a code consisting of 2- or 3-letter abbreviations of the country names, or 2- or 3-digit numbers that represent the country names, that follow the geographical place names.

coupled modes. 1. In fiber optics, a mode that shares energy among one or more other modes, all of which propagate together. *Note:* The distribution of energy among the coupled modes changes with propagation distance. 2. In microwave transmission, a condition where energy is transferred from the fundamental mode to higher order modes. *Note:* Energy transferred to coupled modes is undesirable in usual microwave transmission in a waveguide.

coupler. *See* directional coupler.

coupling. The desirable or undesirable transfer of energy from one medium, such as a metallic wire or an optical fiber, to another like medium, including fortuitous transfer. *Note:* Examples of coupling include capacitive (electrostatic) coupling, inductive (magnetic) coupling, conducted (resistive or hard-wire) coupling, and fiber-optic coupling.

coupling coefficient. A number that expresses the degree of electrical coupling that exists between two circuits. *Note:* The coupling coefficient is calculated as the ratio of the mutual impedance to the square root of the product of the self-impedances of the coupled circuits, all impedances being expressed in the same units.

coupling efficiency. In fiber optics, the efficiency of optical power transfer between two optical components. *Note 1:* The transfer may take place (a) between an active component, such as an LED, and a passive component, such as an optical fiber, or (b) between two passive components such as two optical fibers. *Note 2:* Coupling efficiency is usually expressed as the ratio, converted to percent, of the input power, *i.e.,* the available power from one component, to the power transferred to the other component.

coupling loss. 1. The loss that occurs when energy is transferred from one circuit, circuit element, or medium to another. *Note:* Coupling loss is usually expressed in the same units—such as watts or dB—as in the originating circuit element or medium. 2. In fiber optics, the power loss that occurs when coupling light from one optical device or medium to another.

cover. The technique of concealing or altering the characteristics of communications patterns for the purpose of denying an unauthorized receiver information that would be of value. *Note: Cover* is a process of modulo two addition of a pseudorandom bit stream generated by a cryptographic device with bits from the control message.

coverage. In radio communications, the geographical area within which service from a radio communications facility can be received.

CPE. *Abbreviation for* customer premises equipment.

cpi. *Abbreviation for* characters per inch.

cps. *Abbreviation for* characters per second. *Note: Formerly, abbreviation for* cycles per second, the unit used to express frequency. However, *hertz*, an SI unit, is the proper unit for frequency.

CPU. *Abbreviation for* central processing unit. 1. The portion of a computer that includes circuits controlling the interpretation and execution of instructions. 2. The portion of a computer that executes programmed instructions, performs arithmetic and logical operations on data, and controls input/output functions. *Synonym* central processor. 3. *Abbreviation* *for* communications processor unit. The portion of a digital communications switch that executes programmed instructions, performs arithmetic and logical operations on signals, and controls input/output functions.

CR. *Abbreviation for* channel reliability, circuit reliability.

CRC. *Abbreviation for* cyclic redundancy check.

critical angle. In geometric optics, at a refractive boundary, the smallest angle of incidence at which total internal reflection occurs. *Note 1:* The angle of incidence is measured with respect to the normal at the refractive boundary. *Note 2:* The critical angle is given by

$$\theta_c = \sin^{-1}\left(\frac{n_1}{n_2}\right),$$

where θ_c is the critical angle, n_1 is the refractive index of the less dense medium, and n_2 is the refractive index of the denser medium. *Note 3:* The incident ray is in the denser medium. *Note 4:* If the incident ray is precisely at the critical angle, the refracted ray is tangent to the boundary at the point of incidence.

critical area. An operational area that requires specific environmental control because of the equipment or information contained therein.

critical frequency. 1. In radio propagation by way of the ionosphere, the limiting frequency at or below which a wave component is reflected by, and above which it penetrates through, an ionospheric layer. 2. At vertical incidence, the limiting frequency at or below which incidence, the wave component is reflected by, and above which it penetrates through, an ionospheric layer. *Note:* The existence of the critical frequency is the result of electron limitation, *i.e.,* the inadequacy of the existing number of free electrons to support reflection at higher frequencies.

critical service. *Synonym* essential service.

critical technical load. That part of the total technical power load required for synchronous communications and automatic switching equipment.

critical wavelength. The free-space wavelength that corresponds to the critical frequency. *Note:* The critical wavelength is equal, in meters, to the speed of light (3×10^8 m/s) divided by the critical frequency in hertz.

cross assembler. An assembler that can run symbolic-language input on one type of computer and produce machine-language output for another type of computer.

cross-band radiotelegraph procedure. A radiotelegraph network operational procedure in which calling stations, such as ship stations, call other stations, such as shore stations, using one frequency, and then shift to another frequency to transmit their messages; the called stations answer using a third frequency.

crossbar switch. A switch that has a plurality of vertical paths, a plurality of horizontal paths, and electromagnetical means, *i.e.,* relays, for interconnecting any one of the vertical paths to any one of the horizontal paths.

cross-connect. *Synonym* cross-connection.

cross-connection. Connections between terminal blocks on the two sides of a distribution frame, or between terminals on a terminal block. *Note:* Connections between terminals on the same block are also called *"straps." Synonyms* cross-connect, jumper.

cross coupling. The coupling of a signal from one channel, circuit, or conductor to another, where it is usually considered to be an undesired signal.

crosslink. A data link between two satellites.

cross modulation. Intermodulation caused by the modulation of the carrier of a desired signal by an undesired signal.

cross-office trunk. A trunk that has its terminations within a single facility.

crosspoint. A single element that (a) is in the array of elements that compose a switch and (b) consists of a set of physical or logical contacts that operate together to extend the speech and signaling channels in a switched network.

cross-polarized operation. The operation of two transmitters on the same frequency, but with polarizations in the opposite sense, *e.g.*, plane polarization with one transmitter-receiver pair being vertically polarized and the other pair horizontally polarized.

cross-site link. In a satellite communications system, the signal power and control connections between the components of an Earth station. *Note:* Examples of cross-site links are (a) links between transmitters and antennas and (b) links between control consoles and transmitters.

crosstalk (XT). 1. Undesired capacitive, inductive, or conductive coupling from one circuit, part of a circuit, or channel, to another. 2. Any phenomenon by which a signal transmitted on one circuit or channel of a transmission system creates an undesired effect in another circuit or channel. *Note:* In telephony, crosstalk is usually distinguishable as speech or signaling tones.

crosstalk coupling. The ratio of the power in a disturbing circuit to the induced power in the disturbed circuit observed at specified points of the circuits under specified terminal conditions. *Note:* Crosstalk coupling is usually expressed in dB. *Synonym* crosstalk coupling loss.

crosstalk coupling loss. *Synonym* crosstalk coupling.

cryptanalysis. 1. Operations performed in converting encrypted messages to plain text without initial knowledge of the crypto-algorithm and/or key employed in the encryption. 2. The study of encrypted texts.

CRYPTO. [The] marking or designator identifying COMSEC keying material used to secure or authenticate telecommunications carrying classified or sensitive U.S. Government or U.S. Government-derived information. *Note:* When written in all upper case letters, CRYPTO has the meaning stated above. When written in lower case as a prefix, crypto and crypt are abbreviations for cryptographic.

cryptochannel. A complete system of crypto-communications between two or more holders. The basic unit for naval cryptographic communication. It includes: (a) the cryptographic aids prescribed; (b) the holders thereof; (c) the indicators or other means of identification; (d) the area or areas in which effective; (e) the special purpose, if any, for which provided; and (f) pertinent notes as to distribution, usage, *etc.* A cryptochannel is analogous to a radio circuit.

cryptographic information. All information significantly descriptive of cryptographic techniques and processes or of cryptographic systems and equipment, or their functions and capabilities, and all cryptomaterial.

cryptography. 1. [The] principles, means, and methods for rendering plain information unintelligible and for restoring encrypted information to intelligible form. 2. The branch of cryptology that treats of the principles, means, and methods of designing and using cryptosystems.

crypto key. *Deprecated term. See* key.

cryptologic. Of or pertaining to cryptology.

cryptology. The science that deals with hidden, disguised, or encrypted communications. It embraces communications security and communications intelligence.

cryptomaterial. All material including documents, devices, equipment, and apparatus essential to the encryption, decryption, or authentication of telecommunications. When classified, it is designated CRYPTO and subject to special safeguards.

cryptonet. Stations that hold a specific key for use. *Note:* Activities that hold key for other than use, such as cryptologistic depots, are not cryptonet members for that key. Controlling authorities are *de facto* members of the cryptonets they control.

cryptosecurity. *See* communications security.

cryptosystem. Associated COMSEC items interacting to provide a single means of encryption or decryption.

crystal oscillator (XO). An oscillator in which the frequency is controlled by a piezoelectric crystal. *Note 1:* A crystal oscillator may require controlled temperature because its operating frequency is a function of temperature. *Note 2:* Types of crystal oscillators include voltage-controlled crystal oscillators (VCXO), temperature-compensated crystal oscillators (TCXO), oven-controlled crystal oscillators (OCXO), temperature-compensated-voltage controlled crystal oscillators (TCVCXO), oven-controlled voltage-controlled crystal oscillators (OCVCXO), microcomputer-compensated crystal oscillators (MCXO), and rubidium crystal oscillators (RbXO).

CSMA. *Abbreviation for* carrier sense multiple access.

CSMA/CA. *Abbreviation for* carrier sense multiple access with collision avoidance.

CSMA/CD. *Abbreviation for* carrier sense multiple access with collision detection.

CSU. *Abbreviation for* channel service unit, circuit switching unit, customer service unit.

C2W. *Abbreviation for* command and control warfare.

CTX. *Abbreviation for* Centrex® service.

cursor. A movable, visible mark used to indicate a position of interest on a display surface. *Note:* A cursor may have a controllable shape, such as an underline, a rectangle, or a pointer and usually indicates where the next character or graphic will be entered or revised.

curvature loss. *Synonym* macrobend loss.

curve-fitting compaction. Data compaction accomplished by substituting an analytical expression for the data to be stored or transmitted. *Note:* Examples of curve-fitting compaction are (a) the breaking of a continuous curve into a series of straight line segments and specifying the slope, intercept, and range for each segment and (b) using a mathematical expression, such as a polynomial or a trigonometric function, and a single point on the corresponding curve instead of storing or transmitting the entire graphic curve or a series of points on it.

customer access. In an Integrated Services Digital Network (ISDN), the portion of the ISDN access that a network provider supplies to connect the customer, *i.e.,* subscriber, installation to the network. *Note:* Customer access includes those network elements or portions of elements that extend from the access switch to the network interface.

customer management complex. In network management, a complex that (a) is controlled by a customer and (b) is responsible for, and performs, maintenance for the customer installation.

customer office terminal. 1. Termination equipment that (a) is located on the customer premises and (b) performs a function that may be integrated into the common carrier equipment. *Note:* An example of a customer office terminal is a stand-alone multiplexer located on the customer premises. 2. The digital loop carrier (DLC) multiplexing function that is near the exchange termination (ET) when provided by a stand-alone multiplexer. *Note:* This function may be integrated into the ET.

customer owned and maintained equipment (COAM). *Deprecated term. See* customer premises equipment.

customer premises equipment (CPE). Terminal and associated equipment and inside wiring located at a subscriber's premises and connected with a carrier's communication channel(s) at the demarcation point *("demarc").* *Note 1:* The demarc is a point established in a building or complex to separate customer equipment from telephone company equipment. *Note 2:* Excluded from CPE are over-voltage protection equipment and pay telephones.

customer-provided equipment. *Deprecated term. See* customer premises equipment.

customer service unit (CSU). A device that provides an accessing arrangement at a user location to either switched or point-to-point, data-conditioned circuits at a specifically established data signaling rate. *Note:* A CSU provides local loop equalization, transient protection, isolation, and central office loop-back testing capability.

custom local area signaling service (CLASS). One of an identified group of network-provided enhanced services. *Note:* A CLASS group for a given network usually includes several enhanced service offerings, such as incoming-call identification, call trace, call blocking, automatic return of the most recent incoming call, call redial, and selective forwarding and programming to permit distinctive ringing for incoming calls.

cutback technique. A destructive technique for determining certain optical fiber transmission characteristics, such as attenuation and bandwidth, by (a) performing the desired measurements on a long length of the fiber under test, (b) cutting the fiber under test at a point near the launching end, (c) repeating the measurements on the short length of fiber, and (d) subtracting the results obtained on the short length to determine the results for the residual long length. *Note 1:* The cut should not be made less than 1 meter from the launch end. However, cutting the fiber so close to the launch end (in a multimode fiber) will introduce errors in the measurements because at that point, modal equilibrium conditions have not been established. The errors so introduced will result in conservative results (*i.e.*, higher transmission losses and lower bandwidths) than would be realized under equilibrium conditions. *Note 2:* Several characteristics may be determined using the same test fiber. *Note 3:* A variation of the cutback technique is the substitution method, in which measurements are made on a full length of fiber, and then on a short length of fiber having the same characteristics (core size, numerical aperture), with the results from the short length being subtracted to give the results for the full length.

cutoff attenuator. A waveguide, of adjustable length, which varies the attenuation of signals passing through it.

cutoff frequency. 1. The frequency either above which or below which the output of a circuit, such as a line, amplifier, or filter, is reduced to a specified level. 2. The frequency below which a radio wave fails to penetrate a layer of the ionosphere at the incidence angle required for transmission between two specified points by reflection from the layer.

cutoff mode. The highest order mode that will propagate in a given waveguide at a given frequency.

cutoff wavelength. 1. The wavelength corresponding to the cutoff frequency. 2. In an uncabled single-mode optical fiber, the wavelength greater than which a particular waveguide mode ceases to be a bound mode. *Note 1:* The cutoff wavelength is usually taken to be the wavelength at which the normalized frequency is equal to 2.405. *Note 2*: The *cabled* cutoff wavelength is usually considered to be a more functional parameter because it takes into consideration the effects of cabling the fiber.

cutover. The physical changing of circuits or lines from one configuration to another.

CVSD. *Abbreviation for* continuously variable slope delta modulation.

cw. *Abbreviation for* carrier wave, continuous wave.

CX. *Abbreviation for* composite signaling.

cxr. *Abbreviation for* carrier.

CX signaling. *Synonym* composite signaling.

cyclic distortion. In telegraphy, distortion that (a) is periodic and (b) is not characteristic, not biased, and not fortuitous. *Note:* Causes of cyclic distortion include irregularities in the duration of contact time of the brushes of a transmitter distributor and interference by disturbing alternating currents.

cyclic redundancy check (CRC). An error-detection scheme that (a) uses parity bits generated by polynomial encoding of digital signals, (b) appends those parity bits to the digital signal, and (c) uses decoding algorithms that detect errors in the received digital signal. *Note:* Error correction, if required, may be accomplished through the use of an automatic repeat-request (ARQ) system.

D*. *(Pronounced "D-Star")* *See* specific detectivity.

D-A. *Abbreviation for* digital-to-analog. *See* digital transmission system.

DACS. *Acronym for* digital access and cross-connect system.

DAMA. *Abbreviation for* demand assignment multiple access.

damping. 1. The progressive diminution with time of certain quantities characteristic of a phenomenon. 2. The progressive decay with time in the amplitude of the free oscillations in a circuit.

dark current. The external current that, under specified biasing conditions, flows in a photoconductive detector when there is no incident radiation.

data. Representation of facts, concepts, or instructions in a formalized manner suitable for communication, interpretation, or processing by humans or by automatic means. Any representations such as characters or analog quantities to which meaning is or might be assigned.

data access arrangement. 1. In public switched telephone networks, a single item or group of items at the customer side of the network interface for data transmission purposes, including all equipment that may affect the characteristics of the interface. 2. A data circuit-terminating equipment (DCE) supplied or approved by a common carrier that permits a DCE or data terminal equipment (DTE) to be attached to the common carrier network. *Note:* Data access arrangements are an integral part of all modems built for the public telephone network.

data attribute. A characteristic of a data element such as length, value, or method of representation.

data bank. 1. A set of data related to a given subject and organized in such a way that it can be consulted by users. 2. A data repository accessible by local and remote users. *Note:* A data bank may contain information on single or multiple subjects, may be organized in any rational manner, may contain more than one database, and may be geographically distributed. More than one data bank may be required to build a comprehensive database.

database. 1. A set of data that is required for a specific purpose or is fundamental to a system, project, enterprise, or business. *Note:* A database may consist of one or more data banks and be geographically distributed among several repositories. 2. A formally structured collection of data. *Note:* In automated information systems, the database is manipulated using a database management system.

database engineering. The discipline involving (a) the conception, modeling, and creation, *i.e.,* programming, of a database, (b) data analysis and administration of the database, and (c) database documentation.

database management system (DBMS). A software system that facilitates (a) the creation and maintenance of a database or databases, and (b) the execution of computer programs using the database or databases.

data burst. *Synonym* burst transmission (def. #2).

data bus. A bus used to transfer data within or to and from a processing unit or storage device.

data circuit connection. The interconnection of any combination of links and trunks, on a tandem basis, by means of switching equipment to facilitate information interchange.

data circuit-terminating equipment. *See* DCE.

data collection facility. A facility for gathering and organizing data from a group of sources.

data communication. The transfer of information between functional units by means of data transmission according to a protocol. *Note:* Data are transferred from one or more sources to one or more sinks over one or more data links.

data communication control procedure. A means used to control the orderly communication of information among stations in a data communication network.

data communications control character. *See* control character.

data communications equipment. *Deprecated term. See* DCE.

data compaction. The reduction of the number of data elements, bandwidth, cost, and time for the generation, transmission, and storage of data without loss of information by eliminating unnecessary redundancy, removing irrelevancy, or using special coding. *Note 1:* Examples of data compaction methods are the use of fixed-tolerance bands, variable-tolerance bands, slope-keypoints, sample changes, curve patterns, curve fitting, variable-precision coding, frequency analysis, and probability analysis. *Note 2:* Simply squeezing noncompacted data into a smaller space, for example by increasing packing density or by transferring data on punched cards onto magnetic tape, is not data compaction. *Note 3:* Whereas data compaction reduces the amount of data used to represent a given amount of information, data compression does not.

data compression. 1. Increasing the amount of data that can be stored in a given domain, such as space, time, or frequency, or contained in a given message length. 2. Reducing the amount of storage space required to store a given amount of data, or reducing the length of message required to transfer a given amount of information. *Note 1:* Data compression may be accomplished by simply squeezing a given amount of data into a smaller space, for example, by increasing packing density or by transferring data on punched cards onto magnetic tape. *Note 2:* Data compression does not reduce the amount of data used to represent a given amount of information, whereas data compaction does. Both data compression and data compaction result in the use of fewer data elements for a given amount of information.

data concentrator. A functional unit that permits a common transmission medium to serve more data sources than there are channels currently available within the transmission medium.

data conferencing repeater. A device that enables any one user of a group of users to transmit a message to all other users in that group. *Synonym* technical control hubbing repeater.

data contamination. *Synonym* data corruption.

data corruption. The violation of data integrity. *Synonym* data contamination.

data country code. A 3-digit numerical country identifier that is part of the 14-digit network terminal numbering plan. *Note:* The data country code prescribed numerical designation further constitutes a segment of the overall 14-digit X.121 numbering plan for a CCITT X.25 network.

data-dependent protection. The application of protective data elements to a data stream in such a manner that the composition of the data stream determines the amount or type of protective elements to be added.

data dictionary. 1. A part of a database management system that provides a centralized repository of information about data in a database, such as meaning, relationship to other data. 2. An inventory that describes, defines, and lists all of the data elements that are stored in a database.

data directory. An inventory that specifies the source, location, ownership, usage, and destination of all of the data elements that are stored in a database.

data element. 1. A named unit of data that, in some contexts, is considered indivisible and in other contexts may consist of data items. 2. A named identifier of each of the entities and their attributes that are represented in a database. 3. A basic unit of information built on standard structures having a unique meaning and distinct units or values. 4. In electronic recordkeeping, a combination of characters or bytes referring to one separate item of information, such as name, address, or age.

Data Encryption Standard (DES). [A] cryptographic algorithm for the protection of unclassified computer data and published by the National Institute of Standards and Technology in Federal Information Processing Standard Publication 46-1. *Note:* DES is not approved for protection of national security classified information.

data forwarder. A device that (a) receives data from one data link and retransmits data representing the same information, using proper format and link protocols, to another data link and (b) may forward data between (a) links that are identical, *i.e.,* TADIL B to TADIL B, (b) links that are similar, *i.e.,* TADIL A to TADIL B, or (c) links that are dissimilar, *i.e.,* TADIL A to TADIL J.

datagram. In packet switching, a self-contained packet, independent of other packets, that contains information sufficient for routing from the originating data terminal equipment (DTE) to the destination DTE without relying on prior exchanges between the equipment and the network. *Note:* Unlike virtual call service, when datagrams are sent there are no call establishment or clearing procedures. Thus, the network may not be able to provide protection against loss, duplication, or misdelivery.

data integrity. 1. [The] condition that exists when data is unchanged from its source and has not been accidentally or maliciously modified, altered, or destroyed. 2. The condition in which data are identically maintained during any operation, such as transfer, storage, and retrieval. 3. The preservation of data for their intended use. 4. Relative to specified operations, the *a priori* expectation of data quality.

data item. 1. A named component of a data element; usually the smallest component. 2. A subunit of descriptive information or value classified under a data element. For example the data element "military personnel grade" contains data items such as sergeant, captain, and colonel.

data link. 1. The means of connecting one location to another for the purpose of transmitting and receiving data. 2. An assembly, consisting of parts of two data terminal equipments (DTEs) and the interconnecting data circuit, that is controlled by a link protocol enabling data to be transferred from a data source to a data sink.

data link escape character (DLE). A transmission control character that changes the meaning of a limited number of contiguously following characters or coded representations.

Data Link Layer. *See* Open Systems Interconnection–Reference Model.

data logging. The dating, time-labeling, and recording of data.

data management. The control of data handling operations–such as acquisition, analysis, translation, coding, storage, retrieval, and distribution of data–but not necessarily the generation and use of data.

data management system. *See* database management system.

data medium. The material in or on which one or more characteristics of the material may be used to represent information statically or dynamically. *Note:* Examples of data media are films, compact optical disks, cards, magnetic disks, magnetic drums, and paper.

data mode. In a communications network, the state of data circuit-terminating equipment (DCE) when connected to a communications channel and ready to transmit data, usually digital data. *Note:* When in the data mode, the DCE is not in a talk or dial mode.

data network identification code (DNIC). In the CCITT International X.121 format, the first four digits of the international data number, the three digits that may represent the data country code, and the 1-digit network code, *i.e.*, the network digit.

data numbering plan area (DNPA). In the U.S. implementation of a CCITT X.25 network, the first three digits of a network terminal number (NTN). *Note:* The 10-digit NTN is the specific addressing information for an end-point terminal in an X.25 network.

data phase. The phase of a data call during which data may be transferred between data terminal equipments (DTEs) that are interconnected via the network. *Note:* The data phase of a data call corresponds to the information transfer phase of an information transfer transaction.

data processing. The systematic performance of operations upon data such as handling, merging, sorting, and computing. *Note:* The semantic content of the original data should not be changed. The semantic content of the processed data may be changed. *Synonym* information processing.

data rate. *See* data signaling rate.

data register. *See* register (def. #2).

data scrambler. A device used in digital transmission systems to convert digital signals into a pseudorandom sequence that is free from long strings of simple patterns, such as marks and spaces. *Note:* The data scrambler facilitates timing extraction, reduces the accumulation of jitter, and prevents baseline drift.

data security. [The] protection of data from unauthorized (accidental or intentional) modification, destruction, or disclosure.

data service unit (DSU). 1. A device used for interfacing data terminal equipment (DTE) to the public telephone network. 2. A type of short-haul, synchronous-data line driver, usually installed at a user location, that connects user synchronous equipment over a 4-wire circuit at a preset transmission rate to a servicing central-office. *Note:* This service can be for a point-to-point or multipoint operation in a digital data network.

data set. *Deprecated term. See* DCE.

***__data signaling rate (DSR).__ The aggregate rate at which data pass a point in the transmission path of a data transmission system. *Note 1:* The DSR is usually expressed in bits per second. *Note 2:* The data signaling rate is given by

$$\sum_{i-1}^{m} \frac{\log_2 n_2}{T_i},$$

where m is the number of parallel channels, n_i is the number of significant conditions of the modulation in the I-th channel, and T_i is the unit interval, expressed in seconds, for the I-th channel. *Note 3:* For serial transmission in a single channel, the DSR reduces to $(1/T)\log_2 n$; with a two-condition modulation, *i.e.*, n=2, the DSR is $1/T$. *Note 4:* For parallel transmission with equal unit intervals and equal numbers of significant conditions on each channel, the DSR is $(m/T)\log_2 n$; in the case of a two-condition modulation, this reduces to m/T. *Note 5:* The DSR may be expressed in bauds, in which case, the factor $\log_2 n_i$ in the above summation formula should be deleted when calculating bauds. *Note 6:* In synchronous binary signaling, the DSR in bits per second may be numerically the same as the modulation rate expressed in bauds. Signal processors, such as four-phase modems, cannot change the DSR, but the modulation rate depends on the line modulation scheme, in accordance with Note 4. For example, in a 2400 b/s 4-phase sending modem, the signaling rate is 2400 b/s on the serial input side, but the modulation rate is only 1200 bauds on the 4-phase output side.

data signaling rate transparency. *See* transparency.

data sink. *See* sink.

data source. *See* source.

data station. Data terminal equipment (DTE), data circuit-terminating equipment (DCE), and any intermediate equipment connected at one location. *Note:* The DCE may be connected directly to a data processing system or it may be a part of the data processing system.

data stream. A sequence of digitally encoded signals used to represent information in transmission.

data subscriber terminal equipment. In the DDN, a general purpose terminal device that consists of (a) all the equipment necessary to provide interface functions, perform code conversion, and transform messages on various data media, such as punched cards, magnetic tapes, and paper tapes, to electrical signals for transmission and (b) all the equipment necessary to convert received electrical signals into data stored or recorded on various data media.

data switching exchange (DSE). The equipment installed at a single location to perform switching functions such as circuit switching, message switching, and packet switching.

data terminal equipment. *See* DTE.

data transfer rate. The average number of bits, characters, or blocks per unit time passing between corresponding equipment in a data transmission system.

data transfer request signal. A call control signal transmitted by the data circuit-terminating equipment (DCE) to the data terminal equipment (DTE) indicating that a request signal, originated by a distant DTE, has been received from a distant DCE to exchange data with the station.

data transfer time. The time between (a) the instant at which a user data unit, such as a character, word, block, or message, is made available to a network for transfer by a transmitting data terminal equipment (DTE) and (b) the receipt of that complete data unit by a receiving DTE.

data transmission. The sending of data from one place to another by means of signals over a channel.

data transmission circuit. The transmission media and the intervening equipment used for the transfer of data between data terminal equipments (DTEs). *Note 1:* A data transmission circuit includes any required signal conversion equipment. *Note 2:* A data transmission circuit may transfer information in (a) one direction only, (b) either direction but one way at a time, or (c) both directions simultaneously.

data volatility. Pertaining to the rate of change in the values of stored data over a period of time.

date. An instant in the passage of time, identified with desired precision by a clock and a calendar. *Note:* An example of a date is 23 seconds after 3:14 PM on February 9, 1926. This date might be represented as 1926FEB091514.23.

date-time group (DTG). In a message, a set of characters, usually in a prescribed format, used to express the day of the month, the hour of the day, the minute of the hour, the time zone, and the year. *Note 1:* The DTG is usually placed in the header of the message. *Note 2:* The DTG may be used as a message identifier if it is unique for each message. *Note 3:* The DTG may indicate either the date and time a message was dispatched by a transmitting station or the date and time it was handed into a transmission facility by a user or originator for dispatch.

dating format. The format used to express the time of an event. *Note:* The time of an event on the UTC time scale is given in the following sequence: hour, day, month, year; *e.g.*, 0917 UT, 30 August 1997. The hour is designated by the 24-hour system.

dB. *Abbreviation for* decibel(s). One tenth of the common logarithm of the ratio of relative powers, equal to 0.1 B (bel). *Note 1:* The decibel is the conventional relative power ratio, rather than the bel, for expressing relative powers because the decibel is smaller and therefore more convenient than the bel. The ratio in dB is given by

$$dB = 10 \log_{10}\left(\frac{P_1}{P_2}\right),$$

where P_1 and P_2 are the actual powers. Power ratios may be expressed in terms of voltage and impedance, E and Z, or current and impedance, I and Z, since

$$P = I^2 Z = \frac{E^2}{Z}.$$

Thus dB is also given by

$$dB = 10 \log_{10}\left(\frac{E_1^2/Z_1}{E_2^2/Z_2}\right) = 10 \log_{10}\left(\frac{I_1^2 Z_1}{I_2^2 Z_2}\right).$$

If $Z_1 = Z_2$, these become

$$dB = 20\log_{10}\left(\frac{E_1}{E_2}\right) = 20\log_{10}\left(\frac{I_1}{I_2}\right)$$

Note 2: The dB is used rather than arithmetic ratios or percentages because when circuits are connected in tandem, expressions of power level, in dB, may be arithmetically added and subtracted. For example, in an optical link, if a known amount of optical power, in dBm, is launched into a fiber, and the losses, in dB, of each component (*e.g.*, connectors, splices, and lengths of fiber) are known, the overall link loss may be quickly calculated with simple addition and subtraction.

dBa. *Abbreviation for* decibels adjusted. Weighted absolute noise power, calculated in dB referenced to 3.16 picowatts (–85 dBm), which is 0 dBa. *Note:* The use of F1A-line or HA1-receiver weighting must be indicated in parentheses as required. A one-milliwatt, 1000-Hz tone will read +85 dBa, but the same power as white noise, randomly distributed over a 3-kHz band (nominally 300 to 3300 Hz), will read +82 dBa, due to the frequency weighting. *Synonym* dBrn adjusted.

dBa(F1A). Weighted absolute noise power in dBa, measured by a noise measuring set with F1A-line weighting. *Note:* F1A weighting is no longer used for DOD applications.

dBa(HA1). Weighted noise power in dBa, measured across the receiver of a 302-type or similar subscriber set, by a noise measuring set with HA1-receiver weighting. *Note:* HA1 weighting is no longer used in DOD applications.

dBa0. Noise power in dBa referenced to or measured at zero transmission level point (0TLP), also called a point of zero relative transmission level (0 dBr). *Note:* It is preferred to convert noise readings from dBa to dBa0, as this makes it unnecessary to know or state the relative transmission level at the point of actual measurement.

dBc. *Abbreviation.* dB relative to the carrier power.

dBi. *Abbreviation.* In the expression of antenna gain, the number of decibels of gain of an antenna referenced to the zero dB gain of a free-space isotropic radiator.

dBm. *Abbreviation.* dB referenced to one milliwatt. *Note 1:* dBm is used in communication work as a measure of absolute power values. Zero dBm equals one milliwatt. *Note 2:* In DOD practice, unweighted measurement is normally understood, applicable to a certain bandwidth, which must be stated or implied. *Note 3:* In European practice, psophometric weighting may be implied, as indicated by context; equivalent to dBm0p, which is preferred.

dBm(psoph). Noise power in dBm, measured with psophometric weighting where

$$dBm\ (psoph) = 10 \log_{10} (pWp) - 90$$
$$= dBa - 84,$$

where *pWp* is power in picowatts psophometrically weighted and *dBa* is the weighted noise power in dB referenced to 3.16 picowatts.

DBMS. *Abbreviation for* database management system.

dBmV. *Abbreviation.* dB referenced to one millivolt across 75 ohms. *Note:* This reference is not equivalent to dBm; it is, in fact, 1.33×10^{-5} milliwatts.

dBm0. Power in dBm referred to or measured at a zero transmission level point (0TLP). *Note 1:* A 0TLP is also called a point of zero relative transmission level (0 dBr0). *Note 2:* Some international documents use dBm0 to mean noise power in dBm0p (psophometrically weighted dBm0). In the United States, dBm0 is not so used.

dBm0p. Noise power in dBm0, measured by a psophometer or noise measuring set having psophometric weighting.

dBr. *Abbreviation.* The power ratio, expressed in dB, between any point and a reference point selected as the zero relative transmission level point. *Note:* Any power expressed in dBr does not specify the absolute power. It is a relative measurement only.

dBrn. *Abbreviation.* dB above reference noise. *Note 1:* Weighted noise power in dB is referred to 1.0 picowatt. Thus, 0dBrn = –90 dBm. Use of 144-line, 144-receiver, or C-message weighting, or flat weighting, must be indicated in parentheses as required. *Note 2:* With C-message weighting, a one-milliwatt, 1000-Hz tone will read +90 dBrn, but the same power as white noise, randomly distributed over a 3-kHz band will read approximately +88.5 dBrn (rounded off to +88 dBrn), because of the frequency weighting. *Note 3:* With 144 weightings, a one-milliwatt, 1000-Hz white noise tone will also read +90 dBrn, but the same 3-kHz power will only read +82 dBrn, because of the different frequency weighting.

dBrn adjusted. *Synonym* dBa.

dBrnC. Weighted noise power in dBrn, measured by a noise measuring set with C-message weighting.

dBrnC0. Noise power in dBrnC referred to or measured at a zero transmission level point (0TLP).

dBrn(f_1-f_2). Flat noise power in dBrn, measured over the frequency band between frequencies f_1 and f_2.

dBrn(144-line). Weighted noise power in dBrn, measured by a noise measuring set with 144-line weighting.

dBv. *Abbreviation.* dB relative to 1 volt peak-to-peak. *Note:* The dBv is usually used for television video signal level measurements.

dBW. *Abbreviation.* dB referenced to one watt.

dBx. *Abbreviation.* dB above reference coupling. *Note:* dBx is used to express the amount of crosstalk coupling in telephone circuits. dBx is measured with a noise measuring set.

dc. *Abbreviation for* direct current.

DCA. *Abbreviation for* Defense Communications Agency. *Now* DISA (Defense Information Systems Agency).

DCE. *Abbreviation for* data circuit-terminating equipment. 1. In a data station, the equipment that performs functions, such as signal conversion and coding, at the network end of the line between the data terminal equipment (DTE) and the line, and that may be a separate or an integral part of the DTE or of intermediate equipment. 2. The interfacing equipment that may be required to couple the data terminal equipment (DTE) into a transmission circuit or channel and from a transmission circuit or channel into the DTE. *Synonyms* data communications equipment *(deprecated)*, data set *(deprecated)*.

DCE clear signal. A call control signal transmitted by data circuit-terminating equipment (DCE) to indicate that it is clearing the associated circuit after a call is finished.

DCE waiting signal. A call control signal at the data-circuit-terminating-equipment/data-terminal-equipment (DCE/DTE) interface that indicates that the DCE is ready for another event in the call establishment procedure.

D channel. In ISDN, the 16-kb/s segment of a 144-kb/s, full-duplex subscriber service channel that is subdivided into 2B+D channels, *i.e.*, into two 64-kb/s clear channels and one 16-kb/s channel for the ISDN basic rate. *Note 1:* The D channel is usually used for out-of-band signaling. The two 64-kb/s clear channels are used for subscriber voice and data services. *Note 2:* The D-channel specifications are addressed in the CCITT Recommendation for the Integrated Services Digital Network (ISDN). *Note 3:* The D-channel may be 64 kb/s for the primary rate ISDN service.

dc patch bay. A patch bay in which dc circuits are grouped.

DCS. *Abbreviation for* Defense Communications System.

DDD. *Abbreviation for* direct distance dialing.

DDN. *Abbreviation for* Defense Data Network.

deadlock. 1. Unresolved contention for the use of a system or component. 2. In computer and data processing systems, an error condition such that processing cannot continue because each of two components or processes is waiting for an action or response from the other. [From Weik '89] 3. A permanent condition in which a system cannot continue to function unless some corrective action is taken.

dead sector. In facsimile, the interval between (a) the end of scanning of one object line and (b) the start of scanning of the following line.

dead space. The area, zone, or volume of space that is within the expected range of a radio, radar, or other transmitted signal but in which the signal is not detectable and therefore cannot be received.

debug. To detect, trace, and eliminate mistakes.

deception. *See* electronic deception.

deception repeater. A device that can (a) receive a signal, (b) amplify, delay, or otherwise manipulate the signal, and (c) retransmit it solely for creating deception.

decibel. *See* dB.

decipher. [To] convert enciphered text to the equivalent plain text by means of a cipher system. *Note:* This does not include solution by cryptanalysis.

decision circuit. A circuit that measures the probable value of a signal element and makes an output signal decision based on the value of the input signal and a predetermined criterion or criteria.

decision instant. In the reception of a digital signal, the instant at which a decision is made by a receiving device as to the probable value of a signal condition. *Synonym* selection position.

decode. 1. To convert data by reversing the effect of previous encoding. 2. To interpret a code. 3. [To] convert encoded text into equivalent plain text by means of a code. *Note:* Decoding does not include deriving plain text by cryptanalysis.

decollimation. In a beam with the minimum possible ray divergence or convergence, any mechanism by which rays are caused to diverge or converge from parallelism. *Note 1:* Decollimation may be deliberate for systems reasons, or may be caused by many factors, such as refractive index inhomogeneities, occlusions, scattering, deflection, diffraction, reflection, and refraction. *Note 2:* Decollimation occurs in applications such as radio, radar, sonar, and optical communications.

decrypt. 1. [A] generic term encompassing decode and decypher. 2. To convert encrypted text into its equivalent plain text by means of a cryptosystem. (This does not include solution by cryptanalysis.) *Note:* The term *"decrypt"* covers the meanings of *"decipher"* and *"decode."*

dedicated circuit. A circuit designated for exclusive use by specified users. *Note:* DOD normally considers a dedicated circuit to be between two users only.

dedicated service. In a communications system, a specified set of functions provided to designated users. *Note:* Dedicated service is usually specified in a communications format, such as voice, digital data, facsimile, or video.

deemphasis. In FM transmission, the process of restoring (after detection) the amplitude-vs.-frequency characteristics of the signal.

deeply depressed cladding fiber. An optical fiber construction, usually a single-mode fiber, that has an outer cladding of approximately the same refractive index as the core, and an inner cladding of very low (depressed) refractive index material between them.

deep space. Space at distances from the Earth equal to or greater than 2×10^6 kilometers.

***de facto* standard.** A standard that is widely accepted and used, but lacks formal approval by a recognized standards organization.

default. Pertaining to the pre-defined initial, original, or specific setting, condition, value, or action a system will assume, use, or take in the absence of instructions from the user.

Defense Communications System (DCS). Department of Defense long-haul voice, data, and record traffic system which includes the Defense Data Network, Defense Satellite Communications Systems, and Defense Switched Network.

Defense Data Network (DDN). A component of the Defense Communications System used for switching Department of Defense automated data processing systems.

Defense Switched Network (DSN). A component of the Defense Communications System that handles Department of Defense voice, data, and video communications.

definition. A figure of merit for image quality. *Note:* In an image, definition is usually expressed in terms of the smallest resolvable element, such as lines per inch, or pels per square inch.

deflection. 1. A change in the direction of a traveling particle, usually without loss of particle kinetic energy, representing a change in velocity without a change in the scalar speed of the particle. 2. A change in the direction of a wave, beam, electron, or other entity, such as might be accomplished by an electric or magnetic field. *Note:* If the deflection is caused by a prism (refraction), a mirror (reflection), or optical grating (diffraction), the specific applicable term should be used.

degradation. 1. The deterioration in quality, level, or standard of performance of a functional unit. 2. In communications, a condition in which one or more of the required performance parameters fall outside predetermined limits, resulting in a lower quality of service. *Note:* Degradation is usually categorized as either *"graceful"* or *"catastrophic."*

degraded service state. The condition that exists when one or more of the required service performance parameters fall outside predetermined limits, resulting in a lower quality of service. *Note:* A degraded service state is considered to exist when a specified level of degradation persists for a specified period of time.

degree of coherence. A dimensionless unit, expressed as a ratio, used to indicate the extent of coherence of an electromagnetic wave such as a lightwave. *Note 1:* For lightwaves, the magnitude of the degree of coherence is equal to the visibility, *V*, of the fringes of a two-beam interference test, as given by

$$V = \frac{I_{max} - I_{min}}{I_{max} + I_{min}},$$

where I_{max} is the intensity at a maximum of the interference pattern, and I_{min} is the intensity at a minimum. *Note 2:* Light is considered to be highly coherent when the degree of coherence exceeds 0.88, partially coherent for values less than 0.88 but more than nearly zero values, and incoherent for nearly zero and zero values.

degree of isochronous distortion. In data transmission, the ratio of (a) the absolute value of the maximum measured difference between the actual and the theoretical intervals separating any two significant instants of modulation (or demodulation) to (b) the unit interval. *Note 1:* These instants are not necessarily consecutive. *Note 2:* The degree of isochronous distortion is usually expressed as a percentage. *Note 3:* The result of the measurement should be qualified by an indication of the period, usually limited, of the observation. For a prolonged modulation (or demodulation), it will be appropriate to consider the probability that an assigned value of the degree of distortion will be exceeded.

degree of start-stop distortion. 1. In asynchronous data transmission, the ratio of (a) the absolute value of the maximum measured difference between the actual and theoretical intervals separating any significant instant of modulation (or demodulation) from the significant instant of the start element immediately preceding it to (b) the unit interval. 2. The highest absolute value of individual distortion affecting the significant instants of a start-stop modulation. *Note:* The degree of distortion of a start-stop modulation (or demodulation) is usually expressed as a percentage. Distinction can be made between the degree of late (positive) distortion and the degree of early (negative) distortion.

dehop. To modify a frequency-hopping signal so that it has a constant center frequency.

dejitterizer. A device that reduces jitter in a digital signal. *Note 1:* A dejitterizer usually consists of an elastic buffer in which the signal is temporarily stored and then retransmitted at a rate based on the average rate of the incoming signal. *Note 2:* A dejitterizer is usually ineffective in dealing with low-frequency jitter, such as waiting-time jitter.

delay. 1. The amount of time by which an event is retarded. 2. The time between the instant at which a given event occurs and the instant at which a related aspect of that event occurs. *Note 1:* The events, relationships, and aspects of the entity being delayed must be precisely specified. *Note 2:* Total delay may be demonstrated by the impulse response of a device or system. *Note 3:* In analog systems, total delay is described in terms of the transfer functions in the frequency domain. *Synonym* delay time. **3**. In radar, the electronic delay of the start of the time base used to select a particular segment of the total.

delay distortion. In a waveform consisting of two or more wave components at different frequencies, distortion caused by the difference in arrival times of the frequency components at the output of a transmission system. *Synonyms* time-delay distortion.

delayed-delivery facility. In a communications network, a facility that stores data, destined for delivery to one or more addresses, for delivery at a later time.

delay encoding. The encoding of binary data to form a two-level signal such that (a) a "0" causes no change of signal level unless it is followed by another "0" in which case a transition to the other level takes place at the end of the first bit period; and (b) a "1" causes a transition from one level to the other in the middle of the bit period. *Note:* Delay encoding is used primarily for encoding radio signals because the frequency spectrum of the encoded signal contains less low-frequency energy than a conventional non-return-to-zero (NRZ) signal and less high-frequency energy than a biphase signal.

delay equalizer. A corrective network designed to make the phase delay or envelope delay of a circuit or system substantially constant over a desired frequency range.

delay line. 1. A transmission line, or equivalent device, used to delay a signal. 2. A single-input-channel device, such as a single-input sequential logic element, in which the output channel state at a given instant, *t,* is the same as the input channel state at the instant *t-n*, where n is a number of time units, *i.e.*, the input sequence undergoes a delay of *n* time units, such as *n* femtoseconds, nanoseconds, or microseconds. *Note:* The delay line may have additional taps yielding output channels with values less than *n*.

delay modulation. *See* delay encoding.

delay time. *Synonym* delay (def. #1).

delay working. In telephone switchboard operations, operation intended to ensure fair distribution of the time of one or more lines among groups of call originators. *Note:* An example of delay working is the withdrawing of one or more lines from general use and placing them under the control of a delay operator so that when other operators book call demands on tickets, the tickets can be passed to the delay operator for connection in the order in which they are booked.

deleted bit. A bit not delivered to the intended destination.

deleted block. A block not delivered to the intended destination.

delimiter. 1. A character used to indicate the beginning and end of a character string, *i.e.,* a symbol stream, such as words, groups of words, or frames. 2. A flag that separates and organizes items of data.

delivered block. A successfully transferred block.

delivered overhead bit. A bit that (a) is successfully transferred to a destination user, (b) performs its primary function within the telecommunications system, and (c) does not represent user information.

delivered overhead block. A block that (a) is successfully transferred to a destination user, (b) performs its primary function within the telecommunications system, and (c) does not contain user information bits.

delivery confirmation. Information returned to the originator indicating that a given unit of information has been delivered to the intended addressee(s).

Dellinger effect. An effect—lasting from several minutes to several hours—that causes electromagnetic sky wave signals to disappear rapidly as a result of greatly increased ionization in the ionosphere caused by increased noise from solar storms. *Synonyms:* Dellinger fadeout, Dellinger fading.

Dellinger fadeout. *Synonym* Dellinger effect.

Dellinger fading. *Synonym* Dellinger effect.

delta modulation (DM). Analog-to-digital signal conversion in which (a) the analog signal is approximated with a series of segments, (b) each segment of the approximated signal is compared to the original analog wave to determine the increase or decrease in relative amplitude, (c) the decision process for establishing the state of successive bits is determined by this comparison, and (d) only the change of information is sent, *i.e.,* only an increase or decrease of the signal amplitude from the previous sample is sent whereas a no-change condition causes the modulated signal to remain at the same 0 or 1 state of the previous sample. *Note:* Examples of delta modulation are continuously variable slope delta modulation, delta-sigma modulation, and differential modulation.

delta-sigma modulation. Delta modulation in which the integral of the input signal is encoded rather than the signal itself. *Note:* Delta-sigma modulation may be achieved by preceding a conventional delta-modulation encoder with an integrating network.

demand assignment. An operation in which several users share access to a communications channel on a real-time basis, *i.e.*, a user needing to communicate with another user on the same network requests the required circuit, uses it, and when the call is finished, the circuit is released, making the circuit available to other users. *Note:* Demand assignment is similar to conventional telephone switching, in which common trunks are provided for many users, on a demand basis, through a limited-size trunk group.

demand assignment multiple access (DAMA). In a communications system, a technique for allocating use of bandwidth among multiple users, based on demand. *Note:* DAMA can be implemented in many ways including TDM and FDM.

demand factor. 1. The ratio of (a) the maximum real power consumed by a system to (b) the maximum real power that would be consumed if the entire load connected to the system were to be activated at the same time. *Note:* The maximum real power is usually integrated over a specified time interval, such as 15 or 30 minutes, and is usually expressed in kilowatts. The real power that would be consumed if the entire load connected to the system were to be activated at the same time is obtained by summing the power required by all the connected equipment. This load is expressed in kilowatts if the consumed real power is expressed in kilowatts. 2. The ratio of (a) the maximum power, integrated over a specified time interval, such as 15 or 30 minutes, and usually expressed in kilowatts, consumed by a system, to (b) the maximum volt-amperes, expressed in kilovolt-amperes if the power is expressed in kilowatts, integrated over a time interval of the same duration, though not necessarily during the same interval. *Note:* Charges for electrical power may be based on the demand factor as well as the kilowatt-hours of electrical energy consumed.

demand load. 1. In general, the total power required by a facility. *Note:* The demand load is the sum of the operational load (including any tactical load) and nonoperational demand loads. It is determined by applying the proper demand factor to each of the connected loads and a diversity factor to the sum total. 2. At a communications center, the power required by all automatic switching, synchronous, and terminal equipment (operated simultaneously on-line or in standby), control and keying equipment, plus lighting, ventilation, and air- conditioning equipment required to maintain full continuity of communications. 3. The power required for ventilating equipment, shop lighting, and other support items that may be operated simultaneously with the technical load. 4. The sum of the technical demand and nontechnical demand loads of an operating facility. 5. At a receiver facility, the power required for all receivers and auxiliary equipment that may be operated on prime or spare antennas simultaneously, those in standby condition, multicouplers, control and keying equipment, plus lighting, ventilation, and air conditioning equipment required for full continuity of communications. 6. At a transmitter facility, the power required for all transmitters and auxiliary equipment that may be operated on prime or spare antennas or dummy loads simultaneously, those in standby condition, control and keying equipment, plus lighting, ventilation, and air conditioning equipment required for full continuity of communications.

demand service. In ISDN applications, a telecommunications service that establishes an immediate communication path in response to a user request made through user-network signaling.

demarc. *Acronym for* demarcation point.

demarcation point (demarc). That point at which operational control or ownership of communications facilities changes from one organizational entity to another. *Note:* The demarcation point is usually the interface point between customer-premises equipment and external network service provider equipment. *Synonym* network terminating interface.

democratically synchronized network. A mutually synchronized network in which all clocks in the network are of equal status and exert equal amounts of control on the others, the network operating clock pulse repetition rate being the mean of the natural (uncontrolled) clock pulse repetition rates of the population of clocks.

demodulation. The recovery, from a modulated carrier, of a signal having substantially the same characteristics as the original modulating signal.

demultiplex (DEMUX). *See* demultiplexing.

demultiplexing. The separation of two or more channels previously multiplexed; *i.e.*, the reverse of multiplexing.

deMUX. *Acronym for* demultiplex and demultiplexer.

dense binary code. A binary code in which all possible bit patterns that can be made from a fixed number of bits are used to encode user information but no overhead information. *Note:* Examples of dense binary codes are (a) a pure binary representation for sexadecimal digits using all sixteen possible patterns and (b) an octal representation using all eight patterns. A binary representation of decimal numbers using four binary digits of which only 10 of the possible 16 patterns are used is not a dense binary code. If a binary code is not dense, the unused patterns can be used to detect errors inasmuch as they should only occur if there is an error.

density. In a facsimile system, a measure of the light transmission or reflection properties of an area of an object. *Note 1:* Density is usually expressed as the logarithm to the base 10 of the ratio of incident to transmitted or reflected irradiance. *Note 2:* There are many types of density, such as diffuse, double diffuse, and specular density, each of which will usually have different numerical values for different materials. The relevant type of density depends on the type of optical system, the component materials of the object, and the surface characteristics of the object.

Department of Defense (DOD) master clock. *See* DOD master clock.

departure angle. The angle between the axis of the main lobe of an antenna pattern and the horizontal plane at the transmitting antenna. *Synonym* takeoff angle.

depolarization. 1. Reducing or randomizing the polarization of an electromagnetic wave. *Note:* Depolarization may be caused by transmission through a nonhomogeneous medium or a depolarizer. 2. Prevention of polarization in an electric cell or battery.

depressed-cladding fiber. *Synonym* doubly clad fiber.

depressed-inner-cladding fiber. *Synonym* doubly clad fiber.

dequeue. *Abbreviation for* double-ended queue.

deregulation. A reduction in regulation of (a) tariffs, (b) market entry and exit, and/or (c) facilities in public telecommunication services.

DES. *Abbreviation for* Data Encryption Standard.

descrambler. The inverse of a scrambler. *Note:* The descrambler output is a signal restored to the state that it had when it entered the associated scrambler, provided that no errors have occurred.

desensitation. The reduction of desired signal gain as a result of receiver reaction to an undesired signal. *Note:* The gain reduction is generally due to overload of some portion of the receiver (*e.g.*, the AGC circuitry) resulting in desired signal suppression because the receiver will no longer respond linearly to incremental changes in input voltage.

design margin. The additional performance capability above required standard basic system parameters that may be specified by a system designer to compensate for uncertainties.

design objective (DO). In communications systems, a desired performance characteristic for communications circuits and equipment that is based on engineering analyses, but (a) is not considered feasible to mandate in a standard, or (b) has not been tested. *Note 1:* DOs are used because applicable systems standards are not in existence. *Note 2:* Examples of reasons for designating a performance characteristic as a DO rather than as a standard are (a) it may be bordering on an advancement in the state of the art, (b) the requirement may not have been fully confirmed by measurement or experience with operating circuits, and (c) it may not have been demonstrated that the requirement can be met considering other constraints, such as cost and size. *Note 3:* A DO is sometimes established in a standard for developmental consideration. A DO may also specify a performance characteristic that may be used in the preparation of specifications for development or procurement of new equipment or systems.

despotically synchronized network. A synchronized network in which a unique master clock controls all other clocks in the network.

despun antenna. In a rotating communications satellite, an antenna with a main beam that is continuously redirected with respect to the satellite so that the antenna illuminates a given area on the surface of the Earth, *i.e.*, the footprint does not move with respect to the Earth. *Note:* An antenna may be despun mechanically or electrically.

destination routing. In communications system operations, the routing of messages based on the name of the destination office, the destination user, or the address on the message, *i.e.,* the addressee. [From Weik '89]

destination user. In an information transfer transaction, the user that receives information from the source, *i.e.,* from the originating user.

de-stuffing. The controlled deletion of stuffing bits from a stuffed digital signal, to recover the original signal. *Synonyms* negative justification, negative pulse stuffing.

detection. 1. The recovery of information from an electrical or electromagnetic signal. *Note:* Conventional radio waves are usually detected by heterodyning, *i.e.*, coherent reception/detection. In this method of reception/detection, the received signal is mixed, in some type of nonlinear device, with a signal from a local oscillator, to produce an intermediate frequency, *i.e.*, beat frequency, from which the modulating signal is recovered, *i.e.*, detected. The inherent instabilities of available optical sources have, until relatively recently, prevented practical use of coherent reception/ detection in optical communication receivers. At present, coherent optical receivers, using sophisticated technology, are just beginning to emerge from the laboratory into the field. Virtually all existing optical receivers employ direct detection; that is, the received optical signal impinges directly onto a detector. Direct detection is less sensitive than coherent detection. 2. In tactical operations, the perception of an object of possible military interest but unconfirmed by recognition. 3. In surveillance, the determination and transmission by a surveillance system that an event has occurred.

detectivity. The reciprocal of noise equivalent power.

detector. 1. A device that is responsive to the presence or absence of a stimulus. 2. In an AM radio receiver, a circuit or device that recovers the signal of interest from the modulated wave. *Note:* In FM reception, a circuit called a discriminator is used to convert frequency variations to amplitude variations. 3. In an optical communications receiver, a device that converts the received optical signal to another form. *Note:* Currently, this conversion is from optical to electrical power; however, optical-to-optical techniques are under development.

deterministic routing. 1. In a switched network, switching in which the routes between given pairs of nodes are pre-programmed, *i.e.,* are determined, in advance of transmission. *Note:* The routes used to complete a given call through a network are identified, in advance of transmission, in routing tables maintained in each switch database. The tables assign the trunks that are to be used to reach each switch code, area code, and International Access Prefix (IAP), usually with one or two alternate routes. 2. In a non-switched network, routing in which the routes between given pairs of nodes are determined in advance. *Note:* The routes used to send a given message through a network are identified in advance in routing tables maintained in a database.

deterministic transfer mode. An asynchronous transfer mode in which the maximum information transfer capacity of a telecommunication service is provided throughout a call.

Deutsches Institut für Normung (DIN). Germany's standards-setting organization, equivalent to the American National Standards Institute (ANSI).

deviation ratio. In a frequency modulation system, the ratio of the maximum frequency deviation of the carrier to the maximum modulating frequency of the system under specified conditions.

D4. *See* **channel bank.**

D-4. A framing standard for traditional time-division multiplexing, which standard describes user channels multiplexed onto a trunk that has been segmented (framed) into 24 bytes of 8 bits each. *Note:* The multiplexing function is performed in the D-4 framing structure by interleaving bits of consecutive bytes as they are presented from individual circuits into each D-4 frame.

DFSK. *Abbreviation for* double-frequency shift keying.

diad. *Synonym* dibit.

diagnostic program. 1. A program used to investigate the cause or the nature of conditions or problems within specified elements of a system. 2. A computer program that detects, locates, or identifies a fault in equipment, an error in input data, or an error in a computer program.

dialing. In a communications system, using a device that generates signals for selecting and establishing connections. *Note:* The term *"dialing"* is often used to designate or refer to all calling devices used for inserting data to establish connections.

dial mode. Operation of data circuit-terminating equipment (DCE) so that circuitry, such as data terminal equipment (DTE), associated with call origination is directly connected to a communications channel.

dial pulse. A dc pulse produced by an end instrument that interrupts a steady current at a sequence and rate determined by the selected digit and the operating characteristics of the instrument.

dial pulsing. Pulsing in which a direct-current pulse train is produced by interrupting a steady signal according to a fixed or formatted code for each digit and at a standard pulse repetition rate. *Note:* Dial pulsing originated with rotary mechanical devices integrated into telephone instruments, for the purpose of signaling. Subsequent applications use electronic circuits to generate dial pulses. *Synonym* pulsing.

dial service assistance (DSA). A network-provided service feature, associated with the switching center equipment, in which services, such as directory assistance, call interception, random conferencing, and precedence calling assistance, are rendered by an attendant.

dial signaling. Signaling in which dual tone multifrequency (DTMF) signals or pulse trains are transmitted to a switching center. *Note 1:* Rotary dials produce pulse trains. Keypads may produce either DTMF signals or pulse trains. *Note 2:* Dial signaling traditionally refers to pulse trains only.

dial switching equipment. Switching equipment actuated by electrical impulses generated by a dial or key pulsing arrangement.

dial through. A technique, applicable to access circuits, that permits an outgoing routine call to be dialed by the PBX user after the PBX attendant has established the initial connection.

dial tone. A tone employed in a dial telephone system to indicate to the calling party that the equipment is ready to receive dial or tone pulses.

dial-tone delay. The time between the instant of going off-hook and the instant of receiving a dial tone.

dial-up. 1. A service feature in which a user initiates service on a previously arranged trunk or transfers, without human intervention, from an active trunk to a standby trunk. 2. A service feature that allows a computer terminal to use telephone systems to initiate and effect communications with other computers.

diametral index of cooperation. *Synonym* index of cooperation.

dibit. A group of two bits. *Note:* The four possible states for a dibit are 00, 01, 10, and 11. *Synonym* diad.

dichroic filter. An optical filter that reflects one or more optical bands or wavelengths and transmits others, while maintaining a nearly zero coefficient of absorption for all wavelengths of interest. *Note:* A dichroic filter may be high-pass, low-pass, band-pass, or band rejection.

dichroic mirror. A mirror used to reflect light selectively according to its wavelength.

DID. *Abbreviation for* direct inward dialing.

dielectric. 1. A substance in which an electric field may be maintained with zero or near-zero power dissipation, *i.e.*, the electrical conductivity is zero or near zero. *Note 1:* A dielectric material is an electrical insulator. *Note 2:* In a dielectric, electrons are bound to atoms and molecules, hence there are few free electrons. 2. Pertaining to a substance that has a zero or near zero electrical conductivity.

dielectric filter. *See* interference filter.

dielectric lens. In the radio regime, a lens made of dielectric material that refracts radio waves in the same manner that an optical lens refracts light waves.

dielectric strength. 1. Of an insulating material, the maximum electric field strength that it can withstand intrinsically without breaking down, *i.e.*, without experiencing failure of its insulating properties. *Note:* The theoretical dielectric strength of a material is an intrinsic property of the bulk material and is not dependent on the configuration of the material or the electrodes with which the field is applied. 2. For a given configuration of dielectric material and electrodes, the minimum electric field that produces breakdown. *Note 1:* At breakdown, the electric field frees bound electrons, turning the material into a conductor. *Note 2:* The field strength at which breakdown occurs in a given case is dependent on the respective geometries of the dielectric (insulator) and the electrodes with which the electric field is applied, as well as the rate of increase at which the electric field is applied. *Note 3:* The electric field strength is usually expressed in volts per meter.

dielectric waveguide. A waveguide that consists of a dielectric material surrounded by another dielectric material, such as air, glass, or plastic, with a lower refractive index. *Note 1:* An example of a dielectric waveguide is an optical fiber. *Note 2:* A metallic waveguide filled with a dielectric material is not a dielectric waveguide.

differential encoding. Encoding in which signal significant conditions represent binary data, such as "0" and "1", and are represented as changes to succeeding values rather than with respect to a given reference. *Note:* An example of differential encoding is phase-shift keying (PSK) in which the information is not conveyed by the absolute phase of the signal with respect to a reference, but by the difference between phases of successive symbols, thus eliminating the requirement for a phase reference at the receiver.

differentially coherent phase-shift keying. *See* coherent differential phase-shift keying.

differentially encoded baseband. *Synonym* non-return-to-mark.

differential Manchester encoding. Encoding in which (a) data and clock signals are combined to form a single self-synchronizing data stream, (b) one of the two bits, *i.e.,* "0" or "1", is represented by no transition at the beginning of a pulse period and a transition in either direction at the midpoint of a pulse period, and (c) the other is represented by a transition at the beginning of a pulse period and a transition at the midpoint of the pulse period. *Note:* In differential Manchester encoding, if a "1" is represented by one transition, a "0" is represented by two transitions, and vice versa.

differential mode attenuation. In an optical fiber, the variation in attenuation among the propagating modes.

differential mode delay. In an optical fiber, the variation in propagation delay that occurs because of the different group velocities of different modes. *Synonym* multimode group delay.

differential mode interference. 1. Interference that causes a change in potential of one side of a signal transmission path relative to the other side. 2. Interference resulting from an interference current path coinciding with the signal path.

differential modulation. Modulation in which the choice of the significant condition for any signal element is dependent on the significant condition for the previous signal element. *Note:* An example of differential modulation is delta modulation.

differential phase-shift keying (DPSK). Phase-shift keying that is used for digital transmission in which the phase of the carrier is discretely varied (a) in relation to the phase of the immediately preceding signal element and (b) in accordance with the data being transmitted.

differential pulse-code modulation (DPCM). Pulse-code modulation in which an analog signal is sampled and the difference between the actual value of each sample and its predicted value, derived from the previous sample or samples, is quantized and converted, by encoding, to a digital signal. *Note:* There are several variations of differential pulse-code modulation.

differential quantum efficiency. In an optical source or detector, the slope of the curve relating output quanta to input quanta.

differentiating network. A network, or circuit, that produces an output waveform that is the time derivative of the input waveform. *Note:* Differentiating networks are used in signal processing, such as for producing short timing pulses from square waves.

diffraction. The deviation of an electromagnetic wavefront from the path predicted by geometric optics when the wavefront interacts with, *i.e.*, is restricted by, a physical object such as an opening (aperture) or an edge. *Note:* Diffraction is usually most noticeable for openings of the order of a wavelength. However, diffraction may still be important for apertures many orders of magnitude larger than the wavelength.

diffraction grating. An array of fine, parallel, equally spaced grooves ("rulings") on a reflecting or transparent substrate, which grooves result in diffractive and mutual interference effects that concentrate reflected or transmitted electromagnetic energy in discrete directions, called *"orders,"* or *"spectral orders." Note 1:* The groove dimensions and spacings are on the order of the wavelength in question. In the optical regime, in which the use of diffraction gratings is most common, there are many hundreds, or thousands, of grooves per millimeter. *Note 2:* Order zero corresponds to direct transmission or specular reflection. Higher orders result in deviation of the incident beam from the direction predicted by geometric (ray) optics. With a normal angle of incidence, the angle è, the deviation of the diffracted ray from the direction predicted by geometric optics, is given by

$$\theta = \pm \sin^{-1}\left(\frac{n\lambda}{d}\right),$$

where n is the spectral order, ë is the wavelength, and d is the spacing between corresponding parts of adjacent grooves. *Note 3:* Because the angle of deviation of the diffracted beam is wavelength-dependent, a diffraction grating is dispersive, *i.e.,* it separates the incident beam spatially into its constituent wavelength components, producing a spectrum. *Note 4:* The spectral orders produced by diffraction gratings may overlap, depending on the spectral content of the incident beam and the number of grooves per unit distance on the grating. The higher the spectral order, the greater the overlap into the next-lower order. *Note 5:* By controlling the cross-sectional shape of the grooves, it is possible to concentrate most of the diffracted energy in the order of interest. This technique is called *"blazing."*

diffraction limited. 1. In optics, pertaining to a light beam in which the far-field beam divergence is equal to that predicted by diffraction theory. 2. In focusing optics, pertaining to a light beam in which the impulse response or resolution limit is equal to that predicted by diffraction theory.

diffraction region. In radio propagation, a region outside the line-of-sight region in which radio reception is made possible by the diffraction of the radiated waves.

diffuse reflection. Reflection from a rough or irregular surface which does not maintain the integrity of the incident wavefront. *Contrast with* specular reflection.

digit. A symbol, numeral, or graphic character that represents an integer. *Note 1:* Examples of digits include any one of the decimal characters "0" through "9" and either of the binary characters "0" or "1." *Note 2:* In a given numeration system, the number of allowable different digits, including zero, is always equal to the base (radix).

digital. Characterized by discrete states.

digital access and cross-connect system (DACS). In communications systems, a digital system in which (a) access is performed by T-1 hardware architecture in private and public networks with centralized switching and (b) cross-connection is performed by D3/D4 framing for switching digital-signal-0 (DS-0) channels to other DS-0 channels. *Note:* Modern digital access and cross-connect systems are not limited to the T-carrier system, and may accommodate high data rates such as those of SONET.

digital alphabet. A coded character set in which the characters of an alphabet have a one-to-one relationship with their digitally coded representations.

digital circuit patch bay. A patch bay in which low-level digital data circuits can be patched, monitored, and tested. *Note:* A digital circuit patch bay can be either "D" type (unbalanced) or "K" type (balanced).

digital combining. A method of interfacing digital data signals, in either synchronous or asynchronous mode, without converting the data into a quasi-analog signal.

digital computer. A computer that consists of one or more central processing units (CPUs), that is controlled by internally stored programs, and that stores and processes data in digital form.

digital data. 1. Data represented by discrete values or conditions, as opposed to analog data. 2. Discrete representations of quantized values of variables, *e.g.*, the representation of numbers by digits, perhaps with special characters and the "space" character.

digital error. An inconsistency between the digital signal actually received and the digital signal that should have been received.

digital facsimile equipment. Facsimile equipment that digitally encodes the picture signal, *i.e.*, encodes the baseband signal resulting from scanning the object. *Note:* The facsimile equipment output may be either (a) analog, as defined by CCITT Group 3 protocol, or (b) digital, as defined by CCITT Group 4, STANAG 5000 Type I, and STANAG 5000 Type II protocols.

digital filter. A filter (usually linear), in discrete time, that is normally implemented through digital electronic computation. *Note:* Digital filters differ from continuous time filters only in application. The parameters of digital filters are generally more stable than the parameters of commonly used analog (continuous) filters. Digital filters can be applied as optimal estimators. Commonly used forms are finite impulse response (FIR) and infinite impulse response (IIR).

digital frequency modulation. The transmission of digital data by frequency modulation of a carrier, as in binary frequency-shift keying.

digital group. *See* digroup.

digital loop carrier (DLC). The equipment, including lines, that is used for digital multiplexing of telephone circuits, and that is provided by the network as part of the subscriber access.

digital milliwatt. 1. In digital telephony, a test signal consisting of eight 8-bit words corresponding to one cycle of a sinusoidal signal approximately 1 kHz in frequency and one milliwatt, rms, in power. *Note 1:* The digital milliwatt is stored in ROM. A continuous signal of arbitrary length, *i.e.,* an indefinite number of cycles, may be realized by continually reading out and concatenating the stored information into a data stream to be converted into analog form. *Note 2:* The digital milliwatt is used in lieu of separate test equipment. It has the advantage of being tied in frequency and amplitude to the relatively stable digital clock signal and power (voltage) supply, respectively, that are used by the digital channel bank. 2. A digital signal that is the coded representation of a 0-dBm, 1000-Hertz sine wave.

digital modulation. The process of varying one or more parameters of a carrier wave as a function of two or more finite and discrete states of a signal.

digital multiplexer. A device for combining several digital signals into an aggregate bit stream. *Note:* Digital multiplexing may be implemented by interleaving bits, in rotation, from several digital bit streams either with or without the addition of extra framing, control, or error detection bits.

digital multiplex hierarchy. A hierarchy consisting of an ordered repetition of tandem digital multiplexers that produce signals of successively higher data rates at each level of the hierarchy. *Note 1:* Digital multiplexing hierarchies may be implemented in many different configurations depending on (a) the number of channels desired, (b) the signaling system to be used, and (c) the bit rate allowed by the communications media. *Note 2:* Some currently available digital multiplexers have been designated as Dl-, DS-, or M-series, all of which operate at T-carrier rates. *Note 3:* In the design of digital multiplex hierarchies, care must be exercised to ensure interoperability of the multiplexers used in the hierarchy.

digital network. *See* integrated digital network.

digital phase-locked loop. A phase-locked loop in which the reference signal, the controlled signal, or the controlling signal, or any combination of these, is in digital form.

digital phase modulation. Modulation in which the instantaneous phase of the modulated wave is shifted between a set of predetermined discrete values in accordance with the significant conditions of the modulating signal.

digital primary patch bay. A patch bay that provides (a) the first access of most local user digital circuits in a technical control facility and (b) patching, monitoring, and testing capabilities for both high-level and low-level digital circuits.

digital selective calling (DSC). A synchronous system developed by the International Radio Consultative Committee (CCIR), used to establish contact with a station or group of stations automatically by means of radio. The operational and technical characteristics of this system are contained in CCIR Recommendation 493.

digital signal (DS). A signal in which discrete steps are used to represent information. *Note 1:* In a digital signal, the discrete steps may be further characterized by signal elements, such as significant conditions, significant instants, and transitions. *Note 2:* Digital signals contain m-ary significant conditions.

digital signal 0 (DS0). In T-carrier, a basic digital signaling rate of 64 kb/s, corresponding to the capacity of one voice-frequency-equivalent channel. *Note 1:* The DS0 rate forms the basis for the North American digital multiplex transmission hierarchy. *Note 2:* The DS0 rate may support twenty 2.4-kb/s channels, or ten 4.8-kb/s channels, or five 9.67-kb/s channels, or one 56-kb/s channel, or one 64-kb/s clear channel.

digital signal 1 (DS1). A digital signaling rate of 1.544 Mb/s, corresponding to the North American and Japanese T1 designator.

digital signal 1C (DS1C). A digital signaling rate of 3.152 Mb/s, corresponding to the North American T1C designator.

digital signal 2 (DS2). A digital signaling rate of 6.312 Mb/s, corresponding to the North American and Japanese T2 designator.

digital signal 3 (DS3). 1. A digital signal rate of 44.736 Mb/s, corresponding to the North American T3 designator. 2. A digital signaling rate of 32.064 Mb/s, corresponding to the Japanese T3 designator.

digital signal 4 (DS4). 1. A digital signal rate of 274.176 Mb/s, corresponding to the North American T4 designator. 2. A digital signaling rate of 97.728 Mb/s, corresponding to the Japanese T4 designator.

digital slip. In the reception of a digital data stream, the loss of a bit, or the insertion by the receiver of a bit that was not transmitted, because of a difference in the bit rates of the incoming data stream and the local clock.

digital speech interpolation (DSI). In digital speech transmission, the use of periods of inactivity or constant signal level to increase the transmission efficiency by insertion of additional signals.

digital subscriber line (DSL). In Integrated Services Digital Networks (ISDN), equipment that provides full-duplex service on a single twisted metallic pair at a rate sufficient to support ISDN basic access and additional framing, timing recovery, and operational functions. *Note:* The physical termination of the DSL at the network end is the line termination; the physical termination at the customer end is the network termination.

digital switch. A switch that performs time-division-multiplexed switching of digitized signals. *Note 1:* When used with analog inputs, analog-to-digital and digital-to-analog conversion are required. These functions may be performed by the digital switch. *Note 2:* Implementation is accomplished by the interchange of time slots between input and output ports on a sequential basis under the direction of control systems. The control systems may be automatic, semiautomatic, or manual.

digital switching. Switching in which digitized signals are switched without converting them to or from analog signals.

digital synchronization. Synchronization based on the start of a transmitted digital data unit, such as a bit, character, block, or frame.

digital-to-analog (D-A) converter. A device that converts a digital input signal to an analog output signal carrying equivalent information.

digital transmission group. A group of digitized voice or data channels or both with bit streams that are combined into a single digital bit stream for transmission over communications media. *Note:* Digital transmission groups usually are categorized by their maximum capacity, not by a specific number of channels. However, the maximum digital transmission group capacity must be equal to or greater than the sum of the individual multiplexer input channel capacities.

digital transmission system. A transmission system in which (a) all circuits carry digital signals and (b) the signals are combined into one or more serial bit streams that include all framing and supervisory signals. *Note:* A-D/D-A conversion, if required, is accomplished external to the system.

digital transport. *See* digital transmission system.

digital voice transmission. Transmission of analog voice signals that have been converted into digital signals. *Note:* An example of digital voice transmission is transmission of pulse-code modulated (PCM) analog voice signals.

digitize. To convert an analog signal into a digital signal carrying equivalent information.

digitizer. 1. A device that converts an analog signal into a digital representation of the analog signal. *Note:* A digitizer usually samples the analog signal at a constant sampling rate and encodes each sample into a numeric representation of the amplitude value of the sample. 2. A device that converts the position of a point on a surface into digital coordinate data.

digit time slot. In a digital data stream, a time interval that is allocated to a single digit and that can be uniquely recognized and defined.

digroup. *Abbreviation for* digital group. In telephony, a basic digital multiplexing group. *Note 1:* In the North American and Japanese T-carrier digital hierarchies, each digroup supports 12 PCM voice channels or their equivalent in other services. The DS1 line rate (2 digroups plus overhead bits) is 1.544 Mb/s, supporting 24 voice channels or their equivalent in other services. *Note 2:* In the European hierarchy, each digroup supports 15 PCM channels or their equivalent in other services. The DS1 line rate (2 digroups plus overhead bits) is 2.048 Mb/s, supporting 30 voice channels or their equivalent in other services.

DIN. *Abbreviation for* Deutsches Institut für Normung.

diode laser. *Synonym* injection laser diode.

DIP. *Abbreviation for* dual in-line package.

diplexer. A three-port frequency-dependent device that may be used as a separator or a combiner of signals. *Note:* Duplex transmission through a diplexer is not possible.

diplex operation. The sharing of one common element, such as a single antenna or channel, for transmission or reception of two simultaneous, independent signals on two different frequencies. *Note:* An example of diplex operation is the use of one antenna for two radio transmitters on different frequencies.

dipole antenna. Usually a straight, center-fed, one-half wavelength antenna.

DIP switch. A group of subminiature switches mounted in a package compatible with standard integrated-circuit sockets. *Note:* DIP switches are usually composed of rocker or slide-type switches.

dipulse coding. The coding of "1's" and "0's" in a message in which one full cycle of a square wave, *i.e.,* a positive pulse followed by a negative pulse in the same bit period, is transmitted when the message bit is a "1" and nothing is transmitted when the bit is a "0", or vice versa. *Note:* A dipulse signal can be generated by encoding the data into 50% return-to-zero (RZ) unipolar data and sending the bits through an AND gate with the system clock pulse. This RZ bit stream is then delayed one half-bit period and then added to the undelayed RZ stream. This produces the final dipulse waveform. The dipulse power spectrum is similar to that of the biphase coding power spectrum except dipulse coding produces a pulse-repetition rate equal to the bit rate. [From Weik '89]

Dirac delta function. *Synonym* unit impulse.

direct access. 1. The capability to obtain data from a storage device, or to enter data into a storage device, in a sequence independent of their relative positions by means of addresses that indicate the physical location of the data. 2. Pertaining to the organization and access method that must be used for a storage structure in which locations of records are determined by their keys, without reference to an index or to other records that may have been previously accessed.

direct address. In computing, an address that designates the storage location of an item of data to be treated as an operand.

direct bond. An electrical connection using continuous metal-to-metal contact between the members being joined.

direct-buried cable. A communication cable manufactured or produced for the purpose of burial in direct contact with the earth.

direct call. A facility-handled call in which the network interprets the call request signal as an instruction to establish a connection based on previously designated user information.

direct connect. 1. In computer systems, a permanent communications link that connects directly to a mainframe computer through a terminal controller, usually at binary synchronous (bi-sync) transmission rates. 2. In computer systems, a temporary connection between a microcomputer, *i.e.*, a desktop workstation, and a host bulletin board system or server.

direct coupling. *Synonym* conductive coupling.

direct current signaling (DX signaling). In telephony, a method whereby the signaling circuit E & M leads use the same cable pair(s) as the voice circuit and no filter is required to separate the control signals from the voice transmission.

direct detection. *See* detection (def. #1).

direct dialing service. A service feature that permits a user to place information concerning credit card calls, collect calls, and special billing calls into the public telephone network without operator assistance.

direct distance dialing (DDD). A network-provided service feature in which a call originator may, without operator assistance, call any other user outside the local calling area. *Note 1:* DDD extends beyond the boundaries of national public telephone networks. *Note 2:* DDD requires more digits in the number dialed than are required for calling within the local area.

directed broadcast address. An Internet Protocol address that specifies "all hosts" on a specified network. *Note:* A single copy of a directed broadcast is routed to the specified network, where it is broadcast to all terminals on that network.

directed net. A radio net in which no station other than the net control station may communicate with any other station without first obtaining permission from the net control station. *Note:* A directed net is established by the net control station. The net control station may restore the net to a free net.

direct inward dialing (DID). A service feature that allows inward-directed calls to a PBX to reach a specific PBX extension without human intervention. *Synonym* network inward dialing.

directional antenna. An antenna in which the radiation pattern is not omnidirectional, *i.e.*, a nonisotropic antenna.

directional coupler. A transmission coupling device for separately sampling (through a known coupling loss) either the forward (incident) or the backward (reflected) wave in a transmission line. *Note:* A directional coupler may be used to sample either a forward or backward wave in a transmission line. A unidirectional coupler has available terminals or connections for sampling only one direction of transmission; a bidirectional coupler has available terminals for sampling both directions.

directionalization. The temporary conversion of a portion or all of a two-way trunk group to one-way trunks favoring traffic flowing away from a congested switch. *Note:* Adjacent nodes must cooperate to accomplish directionalization.

direction finding. A procedure for obtaining bearings of radio frequency emitters by using a highly directional antenna and a display unit on an intercept receiver or ancillary equipment.

direction of scanning. In a facsimile transmitting apparatus, the scanning of the plane (developed in the case of a drum transmitter) of the message surface along lines running from right to left commencing at the top so that scanning commences at the top right-hand corner of the surface and finishes at the bottom left-hand corner; this is equivalent to scanning over a right-hand helix on a drum. *Note 1:* The orientation of the message on the scanning plane will depend upon its dimensions and is of no consequence. At the receiving apparatus, scanning takes place from right to left and top to bottom (in the above sense) for "positive" reception and from left to right and top to bottom (in the above sense) for "negative" reception. *Note 2:* This is the CCITT Recommendation for phototelegraphic equipment.

directive gain. 1. Of an antenna, the ratio of (a) 4π times the radiance, *i.e.*, power radiated per unit solid angle (watts per steradian), in a given direction to (b) the total power, *i.e.*, the power radiated to 4π steradians. *Note 1:* The directive gain is usually expressed in dB. *Note 2:* The directive gain is relative to an isotropic antenna. *Note 3:* The power radiated to 4π steradians is the total power radiated by the antenna because 4π steradians constitute an entire sphere. 2. Of an antenna, for a given direction, the ratio of the radiance, *i.e.*, the radiation intensity, produced in the given direction to the average value of the radiance in all directions. *Note 1:* If the direction is not specified, the direction of maximum radiance is assumed. *Note 2:* The directive gain is usually expressed in dB.

directivity pattern. *Synonym* radiation pattern.

direct orbit. For a satellite orbiting the Earth, an orbit in which the projection of the satellite on the equatorial plane revolves about the Earth in the same direction as the rotation of the Earth.

direct outward dialing (DOD). An automated PBX service feature that provides for outgoing calls to be dialed directly from the user terminal. *Synonym* network outward dialing.

direct ray. A ray of electromagnetic radiation that follows the path of least possible propagation time between transmitting and receiving antennas. *Note:* The path of least propagation time is not always the shortest distance path.

direct recording. In facsimile systems, recording in which a visible record is produced, without subsequent processing, in response to received signals.

direct-sequence modulation. In spread-spectrum systems, modulation in which a sequence of binary pulses is used directly to modulate a carrier, usually by phase-shift keying. *Synonym* direct-spread modulation.

direct-sequence spread spectrum. 1. A system (a) for generating spread-spectrum transmissions by phase-modulating a sine wave pseudorandomly with a continuous string of pseudonoise code symbols, each of duration much smaller than a bit and (b) that may be time-gated, where the transmitter is keyed periodically or randomly within a specified time interval. 2. A signal structuring technique utilizing a digital code sequence having a chip rate much higher than the information signal bit rate. Each information bit of a digital signal is transmitted as a pseudorandom sequence of chips.

direct-spread modulation. *Synonym* direct-sequence modulation.

DISA. *Abbreviation for* Defense Information Systems Agency, *formerly* DCA (Defense Communications Agency).

disabling tone. A tone, transmitted over a communications path, used to control equipment. *Note:* An example of a disabling tone is a tone that places an echo suppressor in a nonoperative condition during data transmission over a telephone circuit.

disc. *See* diskette.

DISC. *Abbreviation for* disconnect command.

discone antenna. *See* biconical antenna.

disconnect. In telephony, the disassociation or release of a switched circuit between two stations.

disconnect command (DISC). In Link-Layer protocols, such as high-level data link control (HDLC), synchronous data link control (SDLC), and advanced data communication control procedure (ADCCP), an unnumbered command used to terminate the operational mode previously set.

disconnect signal. In a switched telephone network, a supervisory signal transmitted from one end of a user line or trunk to indicate at the other end that the established connection should be disconnected.

disconnect switch. In a power system, a switch used for closing, opening, or changing the connections in a circuit or system or for purposes of isolation. *Note:* It has no interrupting rating and is intended to be operated only after the circuit has been opened by some other means, such as by a circuit breaker or variable transformer.

discriminator. The part of an FM receiver that extracts the desired signal from an incoming FM wave by changing frequency variations into amplitude variations.

disengagement attempt. An attempt to terminate a telecommunications system access. *Note:* Disengagement attempts may be initiated by a user or the telecommunications system.

disengagement denial. After a disengagement attempt, a failure to terminate the telecommunications system access. *Note:* Disengagement denial is usually caused by excessive delay in the telecommunications system.

disengagement-denial probability. The ratio of disengagement attempts that result in disengagement denial to the total disengagement attempts counted during a measurement period.

disengagement failure. Failure of a disengagement attempt to return a communication system to the idle state, for a given user, within a specified maximum disengagement time.

disengagement originator. The user or functional unit that initiates a disengagement attempt. *Note 1:* A disengagement originator may be the originating user, the destination user, or the communications system. *Note 2:* The communications system may deliberately originate the disengagement because of preemption or inadvertently because of system malfunction.

disengagement phase. In an information transfer transaction, the phase during which successful disengagement occurs. *Note:* The disengagement phase is the third phase of an information transfer transaction.

disengagement request. A control or overhead signal issued by a disengagement originator for the purpose of initiating a disengagement attempt.

disengagement time. 1. The average value of elapsed time between the start of a disengagement attempt for a particular source or destination user and the successful disengagement of that user. 2. Elapsed time between the start of a disengagement attempt and successful disengagement.

diskette. In computer technology, a small disk of flexible plastic, coated with a magnetizable material and enclosed in a protective jacket, used to store digital data. *Note:* A diskette is distinguished from a hard disk by virtue of the fact that it is flexible, and unlike most hard disks, is removable from its drive. *Synonyms* flexible disk, floppy disk.

disk pack. An assembly of magnetic disks that can be removed as a whole from a disk drive together with a container from which the assembly must be separated when operating.

DISNET. *Abbreviation for* Defense Integrated Secure Network.

disparity. In pulse-code modulation (PCM), the digital sum, *i.e.*, the algebraic sum, of a set of signal elements. *Note:* The disparity will be zero and there will be no cumulative or drifting polarization if there are as many positive elements (those that represent 1) as there are negative elements (those that represent 0).

dispersion. Any phenomenon in which the velocity of propagation of an electromagnetic wave is wavelength dependent. *Note 1:* In communication technology, *"dispersion"* is used to describe any process by which an electromagnetic signal propagating in a physical medium is degraded because the various wave components (*i.e.,* frequencies) of the signal have different propagation velocities within the physical medium. *Note 2:* In an optical fiber, there are several significant dispersion effects, such as material dispersion, profile dispersion, and waveguide dispersion, that degrade the signal. *Note 3:* In optical fiber communications, the incorrect terms *"multimode dispersion"* and *"intermodal dispersion"* should not be used as synonyms for the correct term *"multimode distortion."* *Note 4:* In classical optics, *"dispersion"* is used to denote the wavelength dependence of refractive index in matter, (*dn*/*d*ë, where *n* is the refractive index and ë is the wavelength) caused by interaction between the matter and light. *"Dispersion,"* as used in fiber optic communications, should not be confused with *"dispersion"* as used by optical lens designers. *Note 5:* Three types of dispersion, relating to optical fibers, are defined as follows:

material dispersion. In optical fiber communication, the wavelength dependence of the velocity of propagation (of the optical signal) on the bulk material of which the fiber is made. *Note 1:* Because every optical signal has a finite spectral width, material dispersion results in spreading of the signal. *Note 2:* Use of the redundant term *"chromatic dispersion"* is discouraged. *Note 3:* In pure silica, the basic material from which the most common telecommunication-grade fibers are made, material dispersion is minimum at wavelengths in the vicinity of 1.27 μm (slightly longer in practical fibers).

profile dispersion. In an optical fiber, that dispersion attributable to the variation of refractive index contrast with wavelength. Profile dispersion is a function of the profile dispersion parameter.

waveguide dispersion. Dispersion, of importance only in single-mode fibers, caused by the dependence of the phase and group velocities on core radius, numerical aperture, and wavelength. *Note 1:* For circular waveguides, the dependence is on the ratio, a/λ, where a is the core radius and λ is the wavelength. *Note 2:* Practical single-mode fibers are designed so that material dispersion and waveguide dispersion cancel one another at the wavelength of interest.

dispersion coefficient. *See* material dispersion coefficient.

dispersion-limited operation. Operation of a communications link in which signal waveform degradation attributable to the dispersive effects of the communications medium is the dominant mechanism that limits link performance. *Note 1:* The amount of allowable degradation is dependent on the quality of the receiver. *Note 2:* In fiber optic communications, *"dispersion-limited operation"* is often confused with *"distortion-limited operation."*

dispersion-shifted fiber. A single-mode optical fiber that has its minimum-dispersion wavelength shifted, by the addition of dopants, toward its minimum-loss wavelength. *Synonym* EIA Class IVb fiber.

dispersion-unmodified fiber. *Synonym* dispersion-unshifted fiber.

dispersion-unshifted fiber. A single-mode optical fiber that has a nominal zero-dispersion wavelength in the 1.3-μm transmission window. *Synonyms* dispersion-unmodified fiber, EIA Class IVa fiber, nonshifted fiber.

display device. An output unit that gives a visual representation of data.

distance learning. *Synonym* teletraining.

distance measuring equipment (DME). In radio location systems, equipment that ascertains the distance between an interrogator and a transponder.

distance training. *Synonym* teletraining.

distortion. 1. In a system or device, any departure of the output signal waveform from that which should result from the input signal waveform's being operated on by the system's specified, *i.e.*, ideal, transfer function. *Note:* Distortion may result from many mechanisms. Examples include nonlinearities in the transfer function of an active device, such as a vacuum tube, transistor, or operational amplifier. Distortion may also be caused by a passive component such as a coaxial cable or optical fiber, or by inhomogeneities, reflections, *etc.*, in the propagation path. 2. In start-stop teletypewriter signaling, the shifting of the significant instants of the signal pulses from their proper positions relative to the beginning of the start pulse. *Note:* The magnitude of the distortion is expressed in percent of an ideal unit pulse length.

distortion-limited operation. The condition prevailing when distortion of a received signal, rather than its attenuated amplitude (or power), limits performance under stated operational conditions and limits. *Note:* Distortion-limited operation is reached when the system distorts the shape of the waveform beyond specified limits. For linear systems, distortion-limited operation is equivalent to bandwidth-limited operation.

distributed control. Control of a network from multiple points. *Note:* Each point controls a portion of the network, using local information or information transmitted over the network from distant points.

distributed database. 1. A database that is not entirely stored at a single physical location, but rather is dispersed over a network of interconnected computers. 2. A database that is under the control of a central database management system in which storage devices are not all attached to a common processor.

distributed frame-alignment signal. A frame-alignment signal in which the signal elements occupy digit positions that are not consecutive.

distributed network. A network structure in which the network resources, such as switching equipment and processors, are distributed throughout the geographical area being served. *Note:* Network control may be centralized or distributed.

distributed processing. Data processing in which an integrated set of functions is performed within multiple, physically separated devices.

distributed-queue dual-bus (DQDB) [network]. A distributed multi-access network that (a) supports integrated communications using a dual bus and distributed queuing, (b) provides access to local or metropolitan area networks, and (c) supports connectionless data transfer, connection-oriented data transfer, and isochronous communications, such as voice communications.

distributed switching. Switching in which many processor-controlled switching units are distributed, usually close to concentrations of users, and operated in conjunction with a host switch. *Note:* Distributed switching provides improved communications services for concentrations of users remote from the host switch, and reduces the transmission requirements, *i.e.*, the traffic, between such concentrations and the host switch.

distribution. In ISDN applications, the use of broadband transmission of audio or video information to the user without applying any post-production processing to the information. *Note: "Distribution"* is in contrast to *"contribution."*

distribution frame. In communications, a structure with terminations for connecting the permanent wiring of a facility in such a manner that interconnection by cross-connections may readily be made.

distribution-quality television. Television conforming to the NTSC standard, the SECAM standard, the PAL standard, or the PAL-M standard. *Synonym [in CCITT usage]* existing-quality television.

distribution service. In ISDN applications, a telecommunications service that allows one-way flow of information from one point in the network to other points in the network with or without user individual presentation control.

distribution voltage drop. The voltage drop between any two defined points of interest in a power distribution system.

distributor. In data transmission, a device that accepts a data stream from one line and places a sequence of signals, one or more at a time, on several lines, thus performing a spatial multiplexing of the original stream. *Note:* Examples of a distributor are (a) a mechanical unit with input to a rotor and output through many contacts wiped by the rotor and (b) a set of combinational logic circuits, such as a series of AND gates, that are sequentially enabled by a set of pulses and that are all connected to a common bus carrying the input signals.

disturbance voltage. An unwanted voltage induced in a system by natural or man-made sources. *Note:* In telecommunications systems, the disturbance voltage creates currents that limit or interfere with the interchange of information. An example of a disturbance voltage is a voltage that produces (a) false signals in a telephone, (b) noise in a radio receiver, or (c) distortion in a received signal.

diurnal phase shift. The phase shift of electromagnetic signals associated with daily changes in the ionosphere. *Note 1:* The major changes usually occur during the period of time when sunrise or sunset is present at critical points along the path. *Note 2:* Significant phase shifts may occur on paths wherein a reflection area of the path is subject to a large tidal range. *Note 3:* In cable systems, significant phase shifts can be occasioned by diurnal temperature variance.

divergence. *See* beam divergence.

diversity. The property of being made up of two or more different elements, media, or methods. *Note:* In communications, diversity is usually used to provide robustness, reliability, or security.

diversity combiner. A circuit or device for combining two or more signals carrying the same information received via separate paths or channels with the objective of providing a single resultant signal that is superior in quality to any of the contributing signals.

diversity factor. The ratio of the sum of the individual maximum demands of the various parts of a power distribution system to the maximum demand of the whole system. *Note:* The diversity factor is always greater than unity.

diversity gain. In radio communications, the ratio of the signal field strength obtained by diversity combining to the signal strength obtained by a single path. *Note:* Diversity gain is usually expressed in dB.

diversity reception. Radio reception in which a resultant signal is obtained by combining or selecting signals, from two or more independent sources, that have been modulated with identical information-bearing signals, but which may vary in their fading characteristics at any given instant. *Note 1:* Diversity reception is used to minimize the effects of fading. *Note 2:* The amount of received signal improvement when using diversity reception is directly dependent on the independence of the fading characteristics.

diversity transmission. Radio communication using a reception technique in which a resultant signal is obtained by combining signals (a) that originate from two or more independent sources that have been modulated with identical information-bearing signals and (b) that may vary in their transmission characteristics at any given instant. *Note 1:* Diversity transmission and reception are used to obtain reliability and signal improvement by overcoming the effects of fading, outages, and circuit failures. *Note 2:* When using diversity transmission and reception, the amount of received signal improvement depends on the independence of the fading characteristics of the signal as well as circuit outages and failures.

divestiture. The court-ordered separation of the Bell Operating Telephone Companies from AT&T.

D layer. *See* D region, ionosphere.

DLE. *Abbreviation for* data link escape character.

DM. *Abbreviation for* delta modulation.

DO. *Abbreviation for* design objective.

DOD. *Abbreviation for* Department of Defense, direct outward dialing.

DOD master clock. The master clock to which time and frequency measurements for the U.S. Department of Defense are referenced, *i.e.*, are traceable. *Note 1:* The U.S. Naval Observatory master clock is designated as the DOD Master Clock. *Note 2:* The U.S. Naval Observatory master clock is one of the two standard time and frequency references for the U.S. Government in accordance with Federal Standard 1002-A. The other standard time and frequency reference for the U.S. Government is the National Institute of Standards and Technology (NIST) master clock.

domain. 1. The independent variable used to express a function. *Note:* Examples of domains are time, frequency, and space. 2. In distributed networks, all the hardware and software under the control of a specified set of one or more host processors.

domain name server. A server that retains the addresses and routing information for TCP/IP LAN users.

Domain Name System (DNS). The online distributed database system that (a) is used to map human-readable addresses into Internet Protocol (IP) addresses, (b) has servers throughout the Internet to implement hierarchical addressing that allows a site administrator to assign machine names and addresses, (c) supports separate mappings between mail destinations and IP addresses, and (d) uses domain names that (i) consist of a sequence of names, *i.e.,* labels, separated by periods, *i.e.,* dots, (ii) usually are used to name Internet host computers uniquely, (iii) are hierarchical, and (iv) are processed from right to left, such as the host nic.ddn.mil has a name (nic – the Network Information Center), a subdomain (ddn – the Defense Data Network), and a primary domain (mil – the MILNET).

domestic fixed public service. A fixed service, the stations of which are open to public correspondence, for radiocommunications originating and terminating solely at points all of which lie within [. . . the entire United States and certain other geographic areas as specified in 47CFR].

domestic public radio services. The land mobile and domestic fixed public services, the stations of which are open to public correspondence.

dominant mode. In a waveguide that can support more than one propagation mode, the mode that propagates with the minimum degradation, *i.e.*, the mode with the lowest cutoff frequency. *Note:* Designations for the dominant mode are TE_{10} for rectangular waveguides and TE_{11} for circular waveguides.

D1. *See* channel bank.

dopant. An impurity added to an optical medium to change its optical properties. *Note:* Dopants are used in optical fibers to control the refractive index profile and other refractive properties of the fiber.

doppler effect. The change in the observed frequency (or wavelength) of a wave, caused by a time rate of change in the effective path length between the source and the point of observation.

doppler shift. The degree of observed change in frequency (or wavelength) of a wave due to the doppler effect.

double-current transmission. *Synonym* polar direct-current telegraph transmission.

double-ended control. *Synonym* double-ended synchronization.

double-ended queue (dequeue). A queue in which the contents may be changed by adding or removing items at either end.

double-ended synchronization. For two connected exchanges in a communications network, a synchronization control scheme in which the phase error signals used to control the clock at one exchange are derived by comparison with the phase of the incoming digital signal and the phase of the internal clocks at both exchanges. *Synonym* double-ended control.

double-frequency shift keying (DFSK). Frequency-shift keying in which two telegraph signals are multiplexed and transmitted simultaneously by frequency shifting among four frequencies.

double modulation. Modulation in which (a) a subcarrier is modulated with an information-bearing signal and (b) the resulting modulated subcarrier is then used to modulate another carrier that has a higher frequency.

double refraction. *Synonym* birefringence.

double-sideband reduced carrier (DSB-RC) transmission. Transmission in which (a) the frequencies produced by amplitude modulation are symmetrically spaced above and below the carrier and (b) the carrier level is reduced for transmission at a fixed level below that which is provided to the modulator. *Note:* In DSB-RC transmission, the carrier is usually transmitted at a level suitable for use as a reference by the receiver, except for the case in which it is reduced to the minimum practical level, *i.e.*, the carrier is suppressed.

double-sideband suppressed-carrier (DSB-SC) transmission. Transmission in which (a) frequencies produced by amplitude modulation are symmetrically spaced above and below the carrier frequency and (b) the carrier level is reduced to the lowest practical level, ideally completely suppressed. *Note:* DSB-SC transmission is a special case of double-sideband reduced carrier transmission.

double-sideband (DSB) transmission. AM transmission in which both sidebands and the carrier are transmitted.

doubly clad fiber. A single-mode fiber that has two claddings. *Note 1:* Each cladding has a refractive index that is lower than that of the core. Of the two claddings, inner and outer, the inner cladding has the lower refractive index. *Note 2:* A doubly clad fiber has the advantage of very low macrobending losses. It also has two zero-dispersion points, and low dispersion over a much wider wavelength range than a singly clad fiber. *Synonyms* depressed-cladding fiber, depressed-inner-cladding fiber, W-profile fiber (from the fact that a symmetrical plot of its refractive index profile superficially resembles the letter W).

down-converter. A device for performing frequency translation in such a manner that the output frequencies are lower in the spectrum than the input frequencies.

downlink. 1. A data link from a satellite or other spacecraft to a terrestrial terminal. 2. A data link from an airborne platform to a ground-based terminal.

downstream. 1. In communications, the direction of transmission flow from the source toward the sink. 2. With respect to the flow of data in a communications path: at a specified point, the direction toward which data are received later than at the specified point.

downtime. The interval during which a functional unit is inoperable.

DPCM. *Abbreviation for* differential pulse-code modulation.

DPSK. *Abbreviation for* differential phase-shift keying.

DQDB. *Abbreviation for* distributed-queue dual-bus.

D region. That portion of the ionosphere existing approximately 50 to 95 km above the surface of the Earth. *Note:* Attenuation of radio waves, caused by ionospheric free-electron density generated by solar radiation, is pronounced during daylight hours. Because solar radiation is not present at night, ionization ceases, and hence attenuation of radio waves ceases.

drift. A comparatively long-term change in an attribute or value of a system or equipment operational parameter. *Note 1:* The drift should be characterized, such as "diurnal frequency drift" and "output level drift." *Note 2:* Drift is usually undesirable and unidirectional, but may be bidirectional, cyclic, or of such long-term duration and low excursion rate as to be negligible.

drop. 1. In a communications network, the portion of a device directly connected to the internal station facilities, such as toward a switchboard or toward a switching center. 2. The central office side of test jacks. 3. A wire or cable from a pole or cable terminus to a building. 4. To delete, intentionally or unintentionally, or to lose part of a signal, such as dropping bits from a bit stream.

drop and insert. In a multichannel transmission system, a process that diverts (drops) a portion of the multiplexed aggregate signal at an intermediate point, and introduces (inserts) a different signal for subsequent transmission in the same position, *e.g.*, time slot or frequency band, previously occupied by the diverted signal. *Note 1:* Signals not of interest at the drop-and-insert point are not diverted. *Note 2:* The diverted signal may be demodulated or reinserted into another transmission system in the same or another time slot or frequency band. *Note 3:* The time slot or frequency band vacated by the diverted signal need not necessarily be reoccupied by another signal. Likewise, a previously unoccupied time slot or frequency band may be occupied by a signal inserted at the drop-and-insert point.

drop channel operation. Operation in which one or more channels of a multichannel system are terminated, *i.e.*, dropped, at an intermediate point between the end terminals of the system.

dropout. 1. In a communications system, a momentary loss of signal. *Note:* Dropouts are usually caused by noise, propagation anomalies, or system malfunctions. 2. A failure to read properly a binary character from data storage. *Note:* A dropout is usually caused by a defect in the storage medium or by a malfunction of the read mechanism. 3. In magnetic tape, disk, card, or drum systems, a recorded signal with an amplitude less than a predetermined percentage of a reference signal.

drop repeater. In a multichannel communications system, a repeater that has the necessary equipment for the local termination, *i.e.*, the dropping, of one or more channels.

drum factor. 1. In facsimile systems in which drums are used, the ratio of drum length to drum diameter. 2. In facsimile systems in which drums are not used, the ratio of (a) the page width to (b) the page length divided by π.

drum speed. In facsimile systems, the rotation rate of the facsimile transmitter or recorder drum. *Note:* Drum speed is usually expressed in revolutions per minute.

DS. *Abbreviation for* digital signal.

DSA. *Abbreviation for* dial service assistance.

DSA board. A local dial office switchboard at which are handled assistance calls, intercepted calls, and calls from miscellaneous lines and trunks. It may also be employed for handling certain toll calls.

DSB. *Abbreviation for* double sideband. *See* double-sideband transmission.

DSB board. A switchboard of a dial system for completing incoming calls received from manual offices.

DSB-SC. *Abbreviation for* double-sideband suppressed carrier. *See* double-sideband suppressed-carrier transmission.

DSC. *Abbreviation for* digital selective calling.

DSE. *Abbreviation for* data switching exchange.

DSI. *Abbreviation for* digital speech interpolation.

DS0. *Abbreviation for* digital signal 0.

DS1 . . . DS4. *Abbreviations for* digital signal 1. . . digital signal 4.

DS1C. *Abbreviation for* digital signal 1C.

DSN. *Abbreviation for* Defense Switched Network.

DSR. *Abbreviation for* data signaling rate.

D-Star(D*). *Synonym* specific detectivity.

DSU. *Abbreviation for* data service unit.

DTE. *Abbreviation for* data terminal equipment. 1. An end instrument that converts user information into signals for transmission or reconverts the received signals into user information. 2. The functional unit of a data station that serves as a data source or a data sink and provides for the data communication control function to be performed in accordance with link protocol. *Note 1:* The data terminal equipment (DTE) may be a single piece of equipment or an interconnected subsystem of multiple pieces of equipment that perform all the required functions necessary to permit users to communicate. *Note 2:* A user interacts with the DTE, or the DTE may be the user. The DTE interacts with the data circuit-terminating equipment (DCE).

DTE clear signal. A call control signal sent by data terminal equipment (DTE) to initiate clearing.

DTE waiting signal. A call control signal, sent by the data-circuit-terminating-equipment/data-terminal-equipment (DCE/DTE) interface, that indicates that the DTE is waiting for a call control signal from the DCE.

DTMF. *Abbreviation for* dual-tone multifrequency (signaling).

D2. *See* channel bank.

D3. *See* channel bank.

D-type patch bay. A patch bay designed for patching and monitoring of unbalanced data circuits at rates up to 1 Mb/s.

dual access. 1. The connection of a user to two switching centers by separate access lines using a single message routing indicator or telephone number. 2. In satellite communications, the transmission of two carriers simultaneously through a single communication satellite repeater.

dual bus. A pair of parallel buses arranged such that the direction of data flow in one bus is opposite to the direction of data flow in the other bus.

dual diversity. The simultaneous combining of (or selection from) two independently fading signals, so that the resultant signal can be detected through the use of space, frequency, angle, time, or polarization characteristics.

dual homing. The connection of a terminal so that it is served by either of two switching centers. *Note:* In dual homing, a single directory number or a single routing indicator is used.

dual in-line package (DIP). An electronic package with a rectangular housing and a row of pins along each of two opposite sides. *Note:* DIP packages may be used for integrated circuits, or for discrete components, such as resistors or toggle switches. An example of a DIP is a microcircuit package with two rows of seven vertical leads that is specially designed for mounting on a printed circuit board.

dual in-line package switch. *See* DIP switch.

dual-precedence message. A message that contains two precedence designations. *Note:* Usually the higher precedence message is for all action addressees and the lower for all information addressees. [From Weik '89]

dual-tone multifrequency (DTMF) signaling. In telephone systems, multifrequency signaling in which standard set combinations of two specific voice band frequencies, one from a group of four low frequencies and the other from a group of four higher frequencies, are used. *Synonyms* multifrequency pulsing, multifrequency signaling. *Note 1:* DTMF signals, unlike dial pulses, can pass through the entire connection to the destination user, and therefore lend themselves to various schemes for remote control after access, *i.e.*, after the connection is established. *Note 2:* Telephones using DTMF usually have 12 keys. Each key corresponds to a different pair of frequencies. Each pair of frequencies corresponds to one of the ten decimal digits, or to the symbol "#" or "*", the "*" being reserved for special purposes. *Note 3:* The standard signal frequency pairs transmitted by DTMF equipment used by the public exchange carriers are as follows:

Button or Digit	Frequencies (Hz)
1	697/1209
2	697/1336
3	697/1477
4	770/1209
5	770/1336
6	770/1477
7	852/1209
8	852/1336
9	852/1477
0	941/1336
*	941/1209
#	941/1477

Note 4: Tactical telephones have 16 keys, the extra 4 being used for precedence. For DSN (Defense Switched Network) the signal frequency pairs transmitted for

the ten decimal digits and the * and # are the same as those used by the public exchange carriers. The additional four keys, corresponding to four different frequency pairs and the precedence, are as follows:

Button or Key	Frequencies (Hz)
FO (Flash Override)	697/1633
F (Flash)	770/1633
I (Immediate)	852/1633
P (Priority)	941/1633

dual-use access line. A user access line normally used for analog voice communication, but which has special conditioning for use as a digital transmission circuit.

duct. 1. In interfacility cabling, a conduit, which may be direct-earth buried or encased in concrete, used to enclose communications or power cables. *Note:* For maximum resistance to rodent attack, direct-earth-buried conduit should have an outside diameter equal to or greater than 6 cm (2.25 in.). 2. *See* atmospheric duct.

ducting. The propagation of radio waves within an atmospheric duct.

dumb terminal. An asynchronous terminal that (a) does not use a transmission control protocol and (b) sends or receives data sequentially one character at a time. *Note:* Dumb terminals usually handle ASCII characters.

dummy load. A dissipative impedance-matched network, usually used at the end of a transmission line to absorb all incident energy. *Note:* The dummy load usually converts the incident energy to thermal energy.

duobinary signal. A pseudobinary-coded signal in which a "0" *("zero")* bit is represented by a zero-level electric current or voltage; a "1" *("one")* bit is represented by a positive-level current or voltage if the quantity of "0" bits since the last "1" bit is even, and by a negative-level current or voltage if the quantity of "0" bits since the last "1" bit is odd. *Note 1:* Duobinary signals require less bandwidth than NRZ. *Note 2:* Duobinary signaling also permits the detection of some errors without the addition of error-checking bits.

duplex cable. A fiber-optic cable that contains two optical fibers.

duplex circuit. A circuit that permits transmission in both directions. *Note:* For simultaneous two-way transmission, *see* full-duplex circuit.

duplexer. In radar systems, a device that isolates the receiver from the transmitter while permitting them to share a common antenna. *Note 1:* A duplexer must be designed for operation in the frequency band used by the receiver and transmitter, and must be capable of handling the output power of the transmitter. *Note 2:* A duplexer must provide adequate rejection of transmitter noise occurring at the receive frequency, and must be designed to operate at, or less than, the frequency separation between the transmitter and receiver. *Note 3:* A duplexer must provide sufficient isolation to prevent receiver desensitization.

duplex operation. Operating method in which transmission is possible simultaneously, in both directions of a telecommunication channel. *Note 1:* This definition is not limited to radio transmission. *Note 2:* In general, duplex operation and semi-duplex operation require two frequencies in radiocommunication; simplex operation may use either one or two. *Synonyms* full-duplex operation, two-way simultaneous operation.

duty cycle. 1. In an ideal pulse train, *i.e.*, one having rectangular pulses, the ratio of the pulse duration to the pulse period. *Note:* For example, the duty cycle is 0.25 for a pulse train in which the pulse duration is 1 μs and the pulse period is 4 μs. 2. The ratio of (a) the sum of all pulse durations during a specified period of continuous operation to (b) the total specified period of operation. 3. In a continuously variable slope delta (CVSD) modulation converter, the mean proportion of binary "1" digits at the converter output in which each "1" indicates a run of a specified number of consecutive bits of the same polarity in the digital output signal. 4. In a periodic phenomenon, the ratio of the duration of the phenomenon in a given period to the period. *Note:* In a piece of electrical equipment, *e.g.*, an electric motor, the period for which it may be operated without deleterious effects, *e.g.*, from overheating.

dwell time. The period during which a dynamic process remains halted in order that another process may occur.

DX signaling. *Abbreviation for* direct current signaling.

DX signaling unit. A duplex signaling unit that repeats "E" and "M" lead signals into a cable pair(s) via "A" and "B" leads. These signals are transmitted on the same cable pair(s) that transmit(s) the message.

dynamically adaptive routing. In route determination for packet-switched networks, adaptive routing in which an algorithm is used that (a) automatically routes traffic around congested, damaged, or destroyed switches and trunks and (b) allows the system to continue to function over the remaining portions of the network.

dynamicizer. *See* parallel-to-serial conversion.

dynamic range. 1. In a system or device, the ratio of (a) a specified maximum level of a parameter, such as power, current, voltage, or frequency to (b) the minimum detectable value of that parameter. *Note:* The dynamic range is usually expressed in dB. 2. In a transmission system, the ratio of (a) the overload level, *i.e.*, the maximum signal power that the system can tolerate without distortion of the signal, to (b) the noise level of the system. *Note:* The dynamic range of transmission systems is usually expressed in dB. 3. In digital systems or devices, the ratio of maximum and minimum signal levels required to maintain a specified bit error ratio.

dynamic variation. A short-term variation (as opposed to long-term drift) in the characteristics of power delivered to electrical equipment. *Note:* Dynamic variations indicate a short-term departure from steady-state conditions.